Heavy-duty Electric Vehicles

Heavy-duty Electric Vehicles

From Concept to Reality

SHASHANK ARORA
School of Engineering, Aalto University,
Espoo, Finland - 02150

ALIREZA TASHAKORI ABKENAR
MoTeC Research Centre, Victoria, Australia - 3136

SHANTHA GAMINI JAYASINGHE
Australian Maritime College, University of Tasmania,
Tasmania, Australia - 7248

KARI TAMMI
School of Engineering, Aalto University,
Espoo, Finland - 02150

ELSEVIER

Butterworth-Heinemann
An imprint of Elsevier

Butterworth-Heinemann is an imprint of Elsevier
The Boulevard, Langford Lane, Kidlington, Oxford OX5 1GB, United Kingdom
50 Hampshire Street, 5th Floor, Cambridge, MA 02139, United States

Library of Congress Cataloging-in-Publication Data
A catalog record for this book is available from the Library of Congress

British Library Cataloguing-in-Publication Data
A catalogue record for this book is available from the British Library

ISBN: 978-0-12-818126-3

For information on all Butterworth-Heinemann publications
visit our website at https://www.elsevier.com/books-and-journals

Publisher: Matthew Deans
Acquisitions Editor: Carrie Bolger
Editorial Project Manager: Rachel Pomery
Production Project Manager: Manju Thirumalaivasan
Cover Designer: Miles Hitchen

Typeset by SPi Global, India

Contents

Foreword

Development of heavy-duty electric vehicles (EVs), which helps toward decreasing both carbon emissions and road congestions, is expected to be the next step in the quest for electrification of transport. However, in comparison to light-duty EVs, the development process of heavy-duty EVs is different and has its own unique challenges in the technological, financial, and institutional domains. Thus, the success of heavy-duty EV uptake largely depends on finding the right solutions for these challenges.

This book presents a detailed methodology for designing heavy-duty EV powertrains. It describes various subsystems, including EV drivetrain, high-voltage energy storage system, and control systems, in a logical and informative manner. Multiple case studies are also provided to offer better understanding of the subject. Moreover, the authors have used a simple and clear language throughout the book, making both reading and grasping of various underpinning technical concepts very easy for the reader.

In my opinion, this book clearly fills an important knowledge gap on the design of heavy-duty EVs. Having both industrial and academic experience, the authors are well-qualified to comment, discuss, and propose technical solutions for various practical challenges that are associated with the development of heavy-duty EVs.

In summary, this book will be very useful for researchers, professional designers, and decision makers active in this field. I, therefore, strongly recommend this book as a must-read for any person who would like to broaden their knowledge of or make a positive contribution to the field of EVs.

Professor Udaya K. Madawala, PhD
Fellow IEEE Distinguished Lecturer (IEEE Power Electronics Society)
Department of Electrical, Computer & Software Engineering
The University of Auckland, Auckland, New Zealand

Preface

A battery electric city bus can be classified as the first on-road application of an electric powertrain technology that is both profitable and sustainable owing to fixed operating routes and schedules for a city bus, which results in a high utilization rate of the available battery capacity of an electric bus. Regarding freight transportation, trucks accounted for only 2% of the on-road vehicles in the European Union in the year 2020 yet were responsible for more than 22% of the total greenhouse gas emissions in the same year. Electrification of heavy-duty vehicles is crucial for containing the emissions from the road transport sector and for meeting the climate action goals.

Challenges to the electrification of heavy-duty road vehicles come from a lack of clear understanding of the government policies, R&D direction, and uncertainty around the performance of various subsystems in an electric powertrain, for example, durability of the battery pack under a range of conditions or the cost of charging infrastructure and its effect on the electric grid. Finding the right technological solution is the key to the successful integration of electric buses and electric trucks in existing public transportation networks, that is, without any negative impact on their daily performance. To that end, this book presents and discusses a detailed methodology for designing heavy-duty electric vehicle (EV) powertrain, from defining the subsystems and component selection to final integration and vehicle supervisory control system design. This book not only discusses technical aspects of motors, power electronics, batteries, and vehicle dynamics; it also explains various aspects of integrating all these systems in a real application.

The book provides a step-by-step heavy-duty EV concept design and development guide, suitable for engineers and people who are interested in the fast-growing EV industry. To begin with, Chapter 1 provides an overview of various standards and regulations established by different governing bodies to direct the research, design, and development process of EVs. Furthermore, a variety of subsidies announced by different countries and agencies to accelerate the introduction and adoption rate of heavy-duty EVs around the world are also comprehensively discussed in

Chapter 1. The remainder of the book is divided into three main sections: (1) EV Drivetrain; (2) HV (high-voltage) Energy Storage and Charging; and (3) EV Control Systems. Chapter 2 describes various EV drivetrain (electric motor and its inverter drive) configurations and discusses their advantages and disadvantages for heavy-duty EV applications. This is followed by Chapter 3, which offers an explanation on the design process and selection of electric motor and drivetrain components parameters. Key discussed topics in this chapter are EV drivetrain topologies, suitable motor types, and motor-generator specification using load cycle data.

Battery technologies are the core of EV technology. Therefore the energy storage section focuses on aspects that are central to engineering of reliable battery packs such as material selection, manufacturing method, and packaging design. Chapter 4 shares insights on advancements made in materials domain, with an aim to develop next generation Li-ion batteries. It also introduces various advanced manufacturing methods, for example, 3D-printing, the dry electrode method, and cell-to-pack technology, which have the potential to push the development boundaries further. On the other hand, Chapter 5 discusses the creation of a robust electrical connection network and designing of appropriate battery packaging and support systems. Correspondingly, various design rules and elements of high-voltage systems needed to contain electric shock hazards are also described. In line with this, the chapter also presents possible means of limiting vibration transmission to battery packs and increasing their thermal stability. The standard procedure for testing propensity of battery packaging designs to thermal runaway propagation is briefly explained. Last, battery swapping mechanisms are covered. Chapter 6 aims to provide readers with an understanding of the charging and discharging processes and state-of-the-art charging technologies. Conductive charging is introduced as the most common and technically feasible technology for charging of the heavy-duty EVs. Nevertheless, the potential of emerging static and dynamic wireless charging technologies is also discussed. Then, charging levels and standards applied in different countries are introduced. The section closes with an analysis of impacts of EV charging system on the electric grid.

Proper functioning of EV control systems is essential to guarantee a smooth and successful operation of EVs on the road. They provide a fail-safe mode that limits damage to critical systems of the vehicle during a failure event. In view of this, Chapter 7 discusses various EV drivetrain operation modes, and explains their respective torque control strategies. Also, heavy-duty EV drivetrain torque mapping strategies in drive mode and brake mode are presented, and zero torque conditions are explained. Ultimately, specific control parameterizations required in inverter drive for optimal performance of the motor are briefly explained, and the methods to set them are introduced in this chapter. Chapter 8 describes the architecture of the battery management system and provides a comprehensive discussion on its essential functionalities, which include charge balance, temperature regulation, and data storage. Basics of heat generation estimation in batteries are also explained along with the temperature regulation methods. Chapter 9 describes the vehicle control structure for an EV application. It explains the concept of EV supervisory controller internal control layers in detail. After that, it discusses supervisory controller power management strategies for optimal performance of the heavy-duty EVs.

The book finishes by taking a close look at various technical, financial, and institutional challenges that impede the large-scale deployment of heavy-duty EVs in Chapter 10. Some technological changes are being planned to overcome these barriers. A brief introduction about some advancements that are in the pipeline is offered in the end.

Authors have extensive academic and industrial research and development experience in various EV design and development projects and have tried to share their knowledge within the context of heavy-duty vehicles. That being said, they anticipate that shared knowledge is directly transferable to nonroad mobile machinery. It is believed the book will be of particular interest to system designers and application engineers active in this field. It would also help the decision-makers in making an informed choice about the most suitable powertrain for their niche application. Last, it is hoped that the book will inspire the young graduates and engineers to make long-term commitments to the EV technology to build a greener future for humankind.

If your tree gains the fruit of knowledge,
You will bring down the lotus sky.

Naser Khosrow (1004–1088 CE)

CHAPTER 1

Heavy-duty Electric Vehicles and Society

1.1 Introduction

Electromobility is the concept of using "electric powertrain" for transporting people and goods with a view to support sustainable development. According to the United Nations Global Technical Regulations No. 20, "electric powertrain" refers to electric circuits comprising traction motors, rechargeable energy storage system (RESS), power electronic converters, associated connectors and wiring harness, and the coupling system for charging the RESS [1]. The European Union (EU) aims to lower their greenhouse gas (GHG) emissions from the transport sector by 60% by the year 2050 in comparison to the 1990 levels [2]. Electromobility enables improving air quality and reducing dependency on fossil fuels, which in turn allows addressing climate change concerns.

In the United States (US), travel by bus contributes 48% toward the public transportation usage. Average transit bus emits only 0.29 kg of carbon dioxide (CO_2) per passenger mile in comparison to an average single-occupancy private vehicle that emits 0.44 kg of CO_2 per passenger mile. CO_2 emissions from the transit bus drop further down to 0.08 kg per passenger mile when the number of passengers in the bus is such that all the seats are taken [3]. The mass transit system is undoubtedly a cleaner alternative when compared with private vehicles. US federal law defines the mass transit system, also known as public transit, or mass transportation or public transportation system, as "regular, continuing shared-ride surface transportation services that are open to the general public defined by age, disability, or low income" [4]. To realize its complete benefits though, it is essential that the full capacity of the public transit system is utilized. However, public transportation system providers have struggled with decreasing ridership in the past few years owing to factors such as employment and residential decentralization, growing automobile availability, coupled with rising incomes. Therefore, investment in zero-emission or electric buses (e-buses) should be encouraged to isolate the effect of vehicle occupancy on its emission level.

Heavy-duty Electric Vehicles
https://doi.org/10.1016/B978-0-12-818126-3.00002-6

All logistics and supply chain systems depend on goods or freight transportation. Ability to move goods from one site to another allows consumption to occur at a different location from production. Demand is generated by people and businesses that need finished products, supplies, or raw materials; this group is called shippers. Freight transportation service is provided by another group called carriers. Carriers are owners or operators of transport machines, such as trains, trucks, airplanes, and ships. Cargo characteristics and length of haul influences the transportation mode choice made by the shippers. Generally, air freight is used for transporting high-value goods that need to be delivered within a few hours. It is the costliest of all the modes. Railroads and marine transport carry slow-moving, low-value bulk cargo over long distances. In contrast, trucks provide a flexible and rapid transport service for high-value commodities, predominantly in short-haul movements at a slightly higher cost than railways. Trucks also provide the "last-mile" transport, that is, connect shippers to other modes of freight transport. Since the year 2000, emissions from trucks have risen at a rate of 2.2% annually. In the year 2020, trucks accounted for only 2% of the vehicles on-road yet emitted 22% of the total GHG emissions from the road transportation sector in the EU [5]. The strong economy has created a surge in demand for freight transportation causing an increase in trucking activity. The trend is expected to increase, which means that we will struggle to meet the commitments made in Paris Agreement until we decarbonize the road freight or trucks [6]. This chapter provides an overview of various standards and regulations developed to guide the design of electric vehicles (EVs). It also outlines various government policies implemented so far to support the introduction of heavy-duty EVs in some of the major EV markets globally.

1.2 Vehicle classification

Heavy-duty vehicles (HDV) used for road-based transportation can be classified into different streams, as shown in Fig. 1. They are briefly explained in the following text:

1. Length: The easiest way to differentiate between different buses is based on their chassis length. The commonly used categories are as follows:
    - minibuses are buses with a length between 6 and 8 m
    - midibuses have a length between 8 and 10 m
    - standard buses include all buses longer than 10 m

    Some buses with chassis length smaller than 6 m are also available; for example, Volkswagen Microbus.

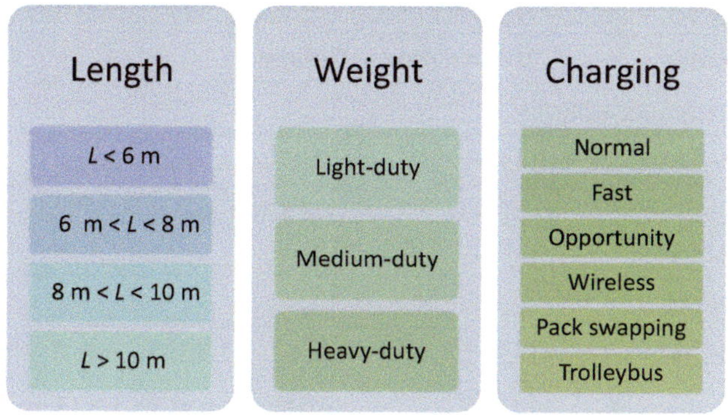

Fig. 1 Classification of electric vehicles.

2. Weight: Gross vehicle weight rating (GVWR) is a safety criterion and represents the maximum weight that a vehicle can operate without overloading. This manufacturer specified limit includes payload capacity in addition to its curb weight.

According to the European Commission's memo—dated May 21, 2014, Category M vehicles, that is, passenger transport vehicles, having more than eight seats are considered HDV. For freight transport or Category N vehicles, heavy-duty classification refers to vehicles weighing more than 3.5 tonnes. Comparatively, there is more granularity in the vehicle classification system defined by the US Environmental Protection Agency (EPA) [7]. Table 1 presents definitions of each of the categories listed by the EPA. However, industrial practice regards vehicles with GVWR up to 6.35 tonnes as light-duty vehicles (LDV), and medium-duty vehicles (MDV) are those with GVWR between 6.351 and 11.793 tonnes. Last, HDV include all vehicles with a GVWR equal to or more than 11.794 tonnes [8].

3. Charging capability: Depending on the battery chemistry employed, e-buses can either allow charging at normal rates, or they can be fast-charged. The former typically employs lithium iron phosphate (LFP), lithium manganese oxide (LMO), nickel cobalt aluminum oxide (NCA), and nickel manganese cobalt oxide (NMC) chemistries, whereas the latter uses lithium titanate oxide (LTO) batteries. Supercapacitors (SCs) can also be used in fast-charging e-buses. The maximum charging rate for normal charging is 1C. In contrast, rates greater than 3C are relatively standard in fast charging [9]. C-rate is a measure of applied or drawn current relative to the rated capacity of the battery. 1C-rate signifies a current that will charge or discharge

Table 1 Summary of vehicle classifications presented in EPA guidelines.

Vehicle category	Abbreviation	Definition
Light-duty passenger vehicles		
Light-duty vehicle	LDV	GVWR < 4.25 tonnes
Medium-duty passenger vehicle	MDPV	GVWR between 4.2505 and 5 tonnes
Light-duty trucks		
Light light-duty trucks	LLDT 1	LVW = 1.875 tonnes
	LLDT 2	LVW more than 1.875 tonnes
Heavy light-duty trucks	HLDT 1	ALVW = 2.875 tonnes
	HLDT 2	ALVW more than 2.875 tonnes
Heavy-duty vehicles		
Light heavy-duty vehicles	HDV 2b	GVWR between 4.2505 and 5 tonnes
	HDV 3	GVWR between 5.0005 and 7 tonnes
	HDV 4	GVWR between 7.0005 and 8 tonnes
	HDV 5	GVWR between 8.0005 and 9.75 tonnes
Medium heavy-duty vehicles	HDV 6	GVWR between 9.7505 and 13 tonnes
	HDV 7	GVWR between 13.0005 and 16.5 tonnes
Heavy heavy-duty vehicles	HDV 8a	GVWR between 16.5005 and 30 tonnes
	HDV 8b	GVWR > 30 tonnes

Please note that the original classification is based on GVWR in pounds. Weight (in pounds) is divided by 2000 to convert it to weight in tonnes. LVW (loaded vehicle weight) = vehicle's curb weight + 150 kg; ALVW (adjusted loaded vehicle weight) = average of vehicle's curb weight and GVWR [7].

the battery in 1 h. Fast-charging e-buses use relatively smaller battery packs because they can be charged quickly at any opportune moment. Others may allow swapping of their battery packs. Furthermore, induction coils can be integrated into the charging infrastructure to charge the battery packs wirelessly. Trolleybuses, that is, e-buses that derive power from overhead cables, are also available in the market. At the end of the year 2017, more than 40% of e-buses in China had a fast-charging option, and only 33% of the e-buses belonged to the normal charging category [10].

1.3 Standards and regulations

A standard is a document that defines a specific procedure or method in detail and sets out requirements for a component, material, item, system, or service. There are national standards and those of international relevance. The international standards for EVs are developed and issued mainly by the ISO (International Organization for Standardization), the IEC (International Electrotechnical Commission), and the SAE (Society of Automotive Engineers). In these organizations, technical committees are charged with the responsibility of preparing different standards. They establish working groups to develop specific portions within a target date.

Please note that standards are "optional" recommendations. In contrast, regulations are mandatory rules and instructions related to performance, safety, and environment compliance [11]. They are the keys that grant original equipment manufacturers (OEMs) access to a marketplace. Therefore, harmonization and homologation of EV standards is essential. Designing EVs and associated infrastructure becomes easy with uniform standards. Motor vehicle industry has partly achieved this goal with the 1958 United Nations Economic Commission for Europe Agreement—comprising more than 130 UN rules. Other 50 countries in addition to the EU region have accepted the agreement.

Fatality of accidents involving HDV is severe owing to their high mass. EVs are typically heavier than their equivalent diesel/petrol-powered counterparts. In view of this, safety of HDV is strictly regulated in regions where road safety actions are encouraged. For example, installation of speed governors is quite common in the EU countries in HDV. Limiting speeds of HDV cause a reduction of 2% in incidents of accident-related injuries. In accordance to the EC Directive 2202/85, speed of commercial vehicles weighing over 3.5 tonnes is limited to 90 km/h. Table 2 lists various national and international regulations that EVs should comply with before they can be approved for sale in a particular region.

Landscape of safety testing in EVs is vast. The UN regulations on RESSs address these at international level. The UN Regulation No. 100 is concerned with the safety of vehicle occupant (driver and passengers) and its mechanical system. Mandatory testing focuses on the mechanical integrity of the system, resistance of the system during use and while being parked, and mechanical shocks. On the other hand, protection and safety of the occupants during and after vehicle crash is covered in UN Regulations 12, 94, and 95. The regulations define factors that affect the safety of vehicle occupants, first responders, and maintenance personnel,

Table 2 National and International Regulations applicable to EV design.

Region	Regulation identifier	Regulation title
Korea	Motor Vehicle Safety Standard, Article 18-2	Safety Tests for High Voltage Systems
	Motor Vehicle Safety Standard, Article 91-4	High Voltage System in Crash Test
Japan	Attachment 110	Technical Standard for Protection of Occupants Against High Voltage in Electric Vehicles and Hybrid Electric Vehicles
	Attachment 111	Technical Standard for Protection of Occupants Against High Voltage After Collision in Electric Vehicles and Hybrid Electric Vehicles
United States	Federal Motor Vehicle Safety Standard 305	Electric-Powered Vehicles: Electrolyte Spillage and Electrical Shock Protection
Canada	Canada Motor Vehicle Safety Standard 305	Electric-Powered Vehicles: Electrolyte Spillage and Electrical Shock Protection
China	GB/T 31498:2015	The Safety Requirement of Electric Vehicle Postcrash
	GB/T 18384.2:2015	Electrically Propelled Road Vehicles-Safety Specifications—Part 2: Vehicle Operational Safety Means and Protection Against Failures
	GB/T 18384.3:2015	Electrically Propelled Road Vehicles-Safety Specifications—Part 3: Protection of Persons Against Electric Shock
	GB/T 31484:2015	Cycle Life Requirements and Test Methods for Traction Battery of Electric Vehicle

	GB/T 31485:2015	Safety Requirements and Test Methods for Traction Battery of Electric Vehicle
	GB/T 31486:2015	Electrical Performance Requirements and Test Methods for Traction Battery of Electric Vehicle
	GB/T 31467.3:2015	Lithium-Ion Traction Battery Pack and System for Electric vehicles—Part 3: Safety Requirements and Test Methods
	GB/T 18384.1:2015	Electrically Propelled Road Vehicles-Safety Specifications—Part 1: On-Board Rechargeable Energy Storage System (RESS)
	GB/T 18384.2:2015	Electrically Propelled Road Vehicles-Safety Specifications—Part 2: Vehicle Operational Safety Means and Protection Against Failures
International	UN Regulation No. 94	Uniform Provisions Concerning the Approval of Vehicles With Regard to the Protection of the Occupants in the Event of a Frontal Collision
	UN Regulation No. 95	Uniform Provisions Concerning the Approval of Vehicles With Regard to the Protection of the Occupants in the Event of a Lateral Collision
	UN Regulation No. 100	Uniform Provisions Concerning the Approval of Vehicles With Regard to Specific Requirements for the Electric Power Train
	UN Regulation No. 12	Concerning the Adoption of Uniform Conditions of Approval and Reciprocal Recognition of Approval for Motor Vehicle Equipment and Parts

for example, electrolyte spillage time. They discuss the importance of mitigating risks of electric shocks and battery fires and other safety components, such as engine controllers, airbags, and antilock braking system. In addition, electromagnetic compliance testing for electronic systems fitted in an EV, such as bluetooth, PTC heaters, immobilizers, GPS, keyless entry, and millimeter-wave radar devices, is covered.

Batteries in an EV generally experience a range of mechanical, electrical, and environmental conditions and may get exposed to

- Vibrations.
- Overcharge/discharge and undercharge/discharge.
- Thermal shock and cycling.
- Mechanical shock and impact.
- Electrical short circuit.
- Fire hazards.

Table 3 presents a list of various standards used by manufacturers for certification of EV battery pack in different regions. International standards such as SAE J2464, SAE J2929, and SAE J2380 address the safety issues at the component level. Safety at vehicle level is covered by the ISO standard 12405. They prescribe test conditions to simulate the working environment of an automotive battery [12]. Criteria for clearing the tests successfully include that the battery/battery pack must be shielded from contact with road debris and must not rupture, leak or spill electrolyte, catch fire, or explode when tested under the prescribed conditions. Following the test procedures defined therein, isolation resistance between the battery terminals and the grounding terminal or battery casing is measured to check whether the mechanical integrity of the pack is intact. Measured values smaller than $100\,\Omega/V$ (working voltage of the DC bus) signify that the pack has been compromised.

A significant difference between the HDV and LDV manufacturers is that to a large extent, the former can also be classified as vehicle integrators. They do not necessarily manufacture complete vehicles, for example, chassis only. The chassis is then handed to a third party for developing the body-in-white structure. The HDV, trucks in particular, can also be customized using different body structures and other design elements to suit different application needs during different phases of their life cycle. Consequently, there can be an abundance of models designed and developed around the same chassis, each of which fits a specific requirement. Customization can involve battery pack orientation, positioning, and fastening. High level of customization can mean incomplete vehicle compliances in some instances. Conditions, implications, and limitations of vehicle-based tests

Table 3 Standards and regulations applicable to battery packs of electric vehicles.

Requirement	International			Europe[a]	USA	India	China [b]	Korea
	Module level	Pack level	Vehicle level					
Mechanical shock	SAE J2464	ISO 12405-3	SAE J2929	UN/ECE R100.02	UL 2580	AIS-048	QC/T 743	
Drop test		SAE J2929			UL 2580		QC/T 743	KMVSS 18-3
Penetration	SAE J2464	SAE J2464			USABC	AIS-048		
Vibration	SAE J2929	ISO 12405-3		UN/ECE R100.02	UL 2580	AIS-048	QC/T 743	
Rollover	SAE J2464	SAE J2929			UL 2580	AIS-048		
Crush/crash	SAE J2464	SAE J2929	ISO 12405-3	UN/ECE R100.02	UL 2580		QC/T 743	KMVSS 91-4
Immersion	SAE J2464	SAE J2929			USABC			KMVSS 18-3
Overcharge/ discharge	SAE J2464	SAE J2929		UN/ECE R100.02	UL 2580	AIS-048		KMVSS 18-3
Short circuit	SAE J2464	ISO 12405-3		UN/ECE R100.02	UL 2580	AIS-048		KMVSS 18-3
Thermal shock	SAE J2464	ISO 12405-3		UN/ECE R100.02	USABC			
Overheat	SAE J2464	SAE J2929		UN/ECE R100.02	USABC			
Fire	SAE J2464	SAE J2929	ISO 12405-3[c]	UN/ECE R100.02	UL 2580			KMVSS 18-3
Emissions	SAE J2464	SAE J2929			USABC			
Flammability	SAE J2464	SAE J2929			USABC			

[a] In addition to the EU region, applicable in countries, including Japan, South Korea, Thailand, Malaysia, Kazakhstan, Australia, New Zealand, Turkey, Azerbaijan, Tunisia, and South Africa.
[b] Applies to Li-ion cells and packs with a rated voltage is a multiplier of 3.6 V, that is, n*3.6 V where, n = number of cells in series connection.
[c] Vehicle body is included.

for HDV must be evaluated carefully. Test load can also become unmanageable very quickly in these cases, which can be avoided by using component-based testing intelligently.

Full vehicle crash tests are conducted in the US in accordance to the Federal Motor Vehicle Safety Standard (FMVSS) 305. Specifically, the frontal barrier, side moving deformable barrier, and the rear moving barrier compliance is tested. Mounting locations are restricted for impact protection. The standard stipulates that postcrash, the battery pack must remain attached to its mounting and secured within its boundaries in the vehicle. More importantly, any of its components should not penetrate the passenger compartment, nor should any electrolyte be spilled in the compartment. Total electrolyte spillage must be contained to less than 5 L. In addition, the electrical isolation must exceed 500 Ω/V (working voltage of the AC bus). Typical recommendations include
Vehicle in-use:

- Protection against direct contact
  - Markings, for example, symbol (black arrow in a triangle with yellow background and black border) on equipment and color coding of high-voltage cables and wires.
  - Physical barriers and access protection, such as enclosures—for charging interfaces, the risk of direct contact depends on their location in the vehicle. Charging interfaces in HDV that are located out of reach of a person standing outside the vehicle are exempt from these requirements.
- Protection against indirect contact
  - Maintaining minimum isolation resistance.
  - Isolation resistance for charge coupling systems.
  - Grounding through conductive charge connectors.
  - Conductive physical barrier bonding.
  - Protection against water effects.

At component level:

- Vibration isolation.
- Overcharge and overdischarge protection.
- Overtemperature protection.
- External short circuit protection.
- Thermal shock and cycling protection.

In postcrash scenarios, the vehicle must comply with any one of the following conditions:

- Low electrical energy—total energy (TE) of the unidirectional single impulse currents is calculated using Eq. (1):

$$\mathrm{TE} = \int_{t_0}^{t_1} (V_b \times I_e)\,dt \quad (1)$$

where V_b is the voltage of high-voltage bus measured between 10 and 60 s of the impact, I_e is the current, t_0 is the time at which the switch is closed, and t_1 refers to the instant at which V_b becomes zero. Calculated value of the TE of the sinusoidal or rectangular current pulses or capacitor discharge from the high-voltage electrical components should be less than 0.2J.

- Absence of high voltage—within 60s of the impact, voltages on the high-voltage bus must be less than or equal to 60V DC or 30V AC (RMS).
- Isolation resistance—should exceed 100 Ω/V DC.
- Physical barriers—resistance less than 0.2 Ω must be maintained between exposed conductive parts of enclosures/physical barriers that are less than 2.5 m apart, are reachable simultaneously, and have at least 0.2 A current flowing through them.

Additionally, there must be no evidence of fire or explosion [1]. Table 4 presents a comprehensive list of various international standards applicable to various aspects of EV design.

1.4 Government policies: subsidies and incentives

Many governments understand the role of nationwide EV deployment in their climate action plans. Several of them have announced various financing plans and incentives to promote the development of e-buses and trucks and support their integration within their transport networks. Some of those significant efforts are discussed in this section.

1.4.1 China

Shared mobility is a cleaner alternative than private vehicles. However, the emission level per passenger varies with vehicle occupancy, which can never be guaranteed. Size of the global e-bus fleet increased by 32% in the year 2018 because China was the first country to understand this correlation and push for electrification of its public transport fleet—421,000 e-buses of approximately 425,000 e-buses operating worldwide are operating in China. China plans to increase its e-bus fleet size to more than 600,000 by the year 2025.

The shift toward electrified public transit system began in the year 2009 with the "Ten Cities, Thousands of Vehicles" incentive program. It was

Table 4 International standards applicable to different subsystems in an EV.

Category	Standard number	Title of the international standard
General	ISO/TR 8713:2019	Electrically Propelled Road Vehicles—Vocabulary
Battery pack—performance and reliability	ISO 12405-4:2018	Electrically Propelled Road Vehicles—Test Specification for Lithium-Ion Traction Battery Packs and Systems—Part 4: Performance Testing
	IEC 62660-2:2010	Secondary Lithium-Ion Cells for the Propulsion of Electric Road Vehicles—Part 2: Reliability and Abuse Testing
	SAE J1766:2014	Recommended Practice for Electric and Hybrid Electric Vehicle Battery Systems Crash Integrity Testing
	SAE J2380:2009	Vibration Testing of Electric Vehicle Batteries
Charging system	IEC 61851-1:2017	Electric Vehicle Conductive Charging System—Part 1: General Requirements
	IEC 61851-21-1:2017	Electric Vehicle Conductive Charging System—Part 21-1 Electric Vehicle On-Board Charger EMC Requirements for Conductive Connection to AC/DC Supply
	IEC 61851-21-2:2018	Electric Vehicle Conductive Charging System—Part 21-2: Electric Vehicle Requirements for Conductive Connection to an AC/DC Supply—EMC Requirements for Off-Board Electric Vehicle Charging Systems
	IEC 61851-23:2014	Electric Vehicles Conductive Charging System—Part 23: DC Electric Vehicle Charging Station
	IEC 61851-24:2014	Electric Vehicles Conductive Charging System—Part 24: Digital Communication Between a DC EV Charging Station and an Electric Vehicle for Control of DC Charging
	IEC 62752:2016	In-Cable Control and Protection Device for Mode 2 Charging of Electric Road Vehicles (IC-CPD)
	SAE J1772:2016	Electric Vehicle and Plug-In Hybrid Electric Vehicle Conductive Charge Coupler

Communication/interface	ISO 15118-1:2019	Road Vehicles—Vehicle to Grid Communication Interface—Part 1: General Information and Use-Case Definition
	ISO 15118-2:2014	Road Vehicles—Vehicle to Grid Communication Interface—Part 2: Network and Application Protocol Requirements
	ISO 15118-3:2015	Road Vehicles—Vehicle to Grid Communication Interface—Part 3: Physical and Data Link Layer Requirements
	ISO 15118-4:2018	Road Vehicles—Vehicle to Grid Communication Interface—Part 4: Network and Application Protocol Conformance Test
	ISO 15118-5:2018	Road Vehicles—Vehicle to Grid Communication Interface—Part 5: Physical Layer and Data Link Layer Conformance Test
	ISO 6722-1:2011	Road Vehicles—60 and 600 V Single-Core Cables—Part 1: Dimensions, Test Methods, and Requirements for Copper Conductor Cables
	ISO 6722-2:2013	Road Vehicles—60 and 600V Single-Core Cables—Part 2: Dimensions Test Methods and Requirements for Aluminum Conductor Cables
	IEC 62196-1:2014	Plugs, Socket-Outlets, Vehicle Connectors, and Vehicle Inlets—Conductive Charging of Electric Vehicles—Part 1: General Requirements
	IEC 62196-2:2011	Plugs, Socket-Outlets, Vehicle Connectors, and Vehicle Inlets—Conductive Charging of Electric Vehicles—Part 2: Dimensional Compatibility and Interchangeability Requirements for AC Pin and Contact-Tube Accessories
	IEC 62196-3:2014	Plugs, Socket-Outlets, and Vehicle Couplers—Conductive Charging of Electric Vehicles—Part 3: Dimensional Compatibility and Interchangeability Requirements for Dedicated DC and Combined AC/DC pin and Contact-Tube Vehicle Couplers

Continued

Table 4 International standards applicable to different subsystems in an EV—cont'd

Category	Standard number	Title of the international standard
Safety	IEC 62660-3:2016	Secondary Lithium-Ion Cells for the Propulsion of Electric Road Vehicles—Part 3: Safety Requirements of Cells and Modules
	SAE J2929:2013	Safety Standard for Electric and Hybrid Vehicle Propulsion Battery Systems Utilizing Lithium-based Rechargeable Cells
	SAE J2344:2010	Guidelines for Electric Vehicle Safety
	SAE J2464:2009	Electric and Hybrid Electric Vehicle Rechargeable Energy Storage System (RESS) Safety and Abuse Testing
	ISO 6469-1:2019	Electrically Propelled Road Vehicles—Safety Specifications—Part 1: Rechargeable energy storage system (RESS)
	ISO 6469-2:2018	Electrically Propelled Road Vehicles—Safety Specifications—Part 2: Vehicle Operational Safety
	ISO 6469-3:2018	Electrically Propelled Road Vehicles—Safety Specifications—Part 3: Electrical Safety
	ISO 6469-4:2015	Electrically Propelled Road Vehicles—Safety Specifications—Part 4: Postcrash Electrical Safety
	ISO 17409:2020	Electrically Propelled Road Vehicles—Conductive Power Transfer—Safety Requirements
	ISO 26262-1:2018	Road Vehicles—Functional Safety—Part 1: Vocabulary
	ISO 26262-2:2018	Road Vehicles—Functional Safety—Part 2: Management of Functional Safety

driven mainly by the widespread urbanization and the goal to limit fossil fuel import. Zhang et al. analyzed the market penetration rate of EVs from 3-years of consumption data of 1.6 million people in Japan. They concluded that increasing purchase subsidy is a more effective tool for encouraging large-scale adoption of EVs than increasing their battery capacity and thereby driving range [13]. However, it was seemingly obvious to the Chinese policymakers from the beginning as comprehensive packages of stringent regulations, fiscal and tax policies, coupled with generous subsidies, for EVs were announced at regular intervals [14]. The main initiatives are briefly described here:

1. 2009–12: The Central government launches the "Ten cities, Thousands of Vehicles" program for 3 years [15]. Only the e-buses with chassis length longer than 10m receive a fixed purchase subsidy of 64,000 euros. In comparison, fuel cell buses of the same length get 77,000 euros rebate. The subsidy is transferred directly to the manufacturer who can then transfer it as a discount to the client.
2. 2013–15: The Central government announces the second phase of the subsidy program. Buses excluded in Phase 1, that is, e-buses smaller than 10m, also get subsidized. Depending on the e-bus size, an amount of 38,000–64,000 euros is available as direct subsidy from the Central government. A minimum all-electric range (AER) of 150km is needed for the e-bus to qualify for the subsidy. Fast-charging e-buses exempt from this condition.

   Complimentary rebate from local governments, at the same level as the national standard, is also available. In addition, annual funding of 2.5–15 million euros, depending on the number of e-buses registered in the region, is available from the central government to different cities for building the charging infrastructure for the e-buses.
3. 2015–20: The Central government announces operational subsidy of up to 10,000 euros (actual amount depends on the size of the fleet) for transit operators to cover operation costs for heavy-duty e-bus fleets. Operators of medium-duty buses and light-duty buses receive 8000 and 6000 euros, respectively, to cover their operational costs. Total subsidy available for each bus is represented by Eq. (2):

   $$S_t = S_c + S_l + S_o \tag{2}$$

   where S_t is the total subsidy amount, S_c is the subsidy received from the central government, S_l is the subsidy awarded by the local governing body, and S_o is the operational subsidy.

4. 2016–20: The third phase of subsidy for e-buses was announced under the "13th Five Year Plan of Urban Transportation" by the Chinese Central government. The objective is to have more than 35% e-buses in the public transportation fleet by the year 2020. Energy consumption rate per load mass (E_{kg}) is added to the subsidy criteria in addition to bus length and AER. Lower the E_{kg}, higher the subsidy amount. Fiscal subsidy from 15,000 to 64,000 euros is available to standard e-buses of 10–12 m length. Subsidy coefficients of 0.2, 0.5, 0.8, and 1.2 applied to e-buses smaller than 6 m, minibuses, midibuses, and e-buses longer than 12 m, respectively. Double decker e-buses are also subsidized at 1.2 times the standard e-buses. A secondary coefficient is multiplied to account for the tonnage of the e-bus. Final subsidy for light-duty, medium-duty, and heavy-duty buses is determined using 0.5, 0.8, and 1.2 as coefficients, respectively. Subsidy for fuel cell buses remains fixed at 64,000 euros [16].

   Techno-economic analysis of the Chinese market revealed that the e-buses were oversubsidized in the year 2016, and thus their market capture rate was exceptional. Unprecedented and unchecked growth curbed technological advancements of the electric drivetrain because the OEMs did not feel the need to innovate. Consequently, it affected sustainable industrial development negatively [17]. The Chinese government updated the third subsidy plan to correct this, and a multicriteria assessment system was announced in the year 2017; hereon referred to as the fourth subsidy phase. Expectations were to introduce a more scientific incentive disbursement scheme and to promote a healthy competition between various OEMs that would significantly alter the technology roadmap for the e-bus industry.

5. 2017–20: The fourth phase of the national subsidy disbursal scheme favors superior battery technology. Battery performance is used for determining the subsidy amount under the revised national policy. Specific energy density of the e-bus battery pack is established as one of the core criteria for determining the subsidy for normal charging e-bus models, and high-energy dense battery packs are subsidized more through multiple discount coefficients. E_{kg} should be less than 0.24 Wh/km kg for the model to qualify for the subsidy. Maximum permissible charging rates attract further discounts to encourage adoption of e-buses with fast-charging option. Table 5 presents the subsidy coefficients and accompanying criteria for evaluating the subsidy level for different e-buses under the fourth subsidy plan [9].

Table 5 Subsidy evaluation criteria for e-buses in the fourth subsidy plan.

Model	Subsidy (RMB/kWh)	Subsidy coefficient		Max. national subsidy (thousand RMB)	
Normal charge	1800	$85 \leq E_s \leq 95$	0.8	$6 < L \leq 8$ m	90
		$95 \leq E_s \leq 115$	1.0	$8 < L \leq 10$ m	200
		$E_s \geq 115$	1.2	$L > 10$ m	300
Fast charge	3000	$3 \leq C_{max} \leq 5$	0.8	$6 < L \leq 8$ m	60
		$5 \leq C_{max} \leq 15$	1.0	$8 < L \leq 10$ m	120
		$C_{max} \geq 15$	1.4	$L > 10$ m	200

E_s and C_{max} are the specific energy density and the maximum permissible charging rate for the chemistry of the installed battery pack, respectively. Please note the maximum subsidy awarded by the local governments is capped at 50% of the national subsidy, i.e., Max. $S_l \leq 0.5\ S_c$

Table 6 Features of the subsidy plan launched by the Chinese national government for accelerating deployment of e-buses in China.

Stage		Phase 1	Phase 2	Phase 3	Phase 4
Coverage period		2009–12	2013–15	2016	2017–20
Models covered	Light-duty			$L \leq 6$	
			$6 \leq L < 8$	$6 < L \leq 8$	$6 < L \leq 8$
	Medium-duty		$8 \leq L < 10$	$8 < L \leq 10$	$8 < L \leq 10$
	Heavy-duty	$L \geq 10$	$L \geq 10$	$L > 10$	$L > 10$
Eligibility criteria					
AER required		×	✓	✓	✓
E_{kg} required		×	×	✓	✓
Battery performance required		×	×	×	✓
Additional benefits					
Subsidy for fast-charging bus		×	✓	✓	✓
Operational subsidy		×	×	✓	✓

Key features of the four national subsidy plans are presented in Table 6.

China's electrification policy has been quite different from that of the rest of the world. It is not just adding e-buses to its fleet, but it is transforming it with the help of local e-bus manufacturers. Their futuristic policy has boosted the local economy significantly and has propelled their manufacturing sector to a new height. Multiple local companies have now become global brands. Moreover, China is developing several new cities. The newness of the cities means the electrified public transport arrives first, and the transit agencies do not have to design an integration plan to get the buses with charging requirements absorbed into the established lines. It is hoped that other countries can replicate this staged transition.

1.4.2 United States

China has a top-down approach to its fleet electrification plan—establish national mandate, encourage EV policy competition among its cities, and incentivize manufacturers for developing e-buses and other heavy-duty EVs. In contrast, there is no definitive industrial policy in the US to promote local manufacturing of e-buses.

Procurement strategy for new buses by many of the US municipalities is based on the end-of-life of the existing bus. Total US public transport fleet size is 65,000 buses. An average bus has a useful life of 12 years, which means that approximately, 5000 buses only can be replaced with new buses each year. High costs and technological limitations have so far restricted proliferation of e-bus fleet in the US. As a result, only 300 e-buses were operational in the US at the end of the year 2017, that is, market penetration for e-buses was less than 0.5%. Recent improvements in battery technology are making them reconsider their position, and 9% of all the US transit agencies have placed orders for e-buses. By the year 2025, the US public transport fleet is expected to have 5000 e-buses.

Majority of the orders are from the metropolitan regions that are affected by poor air quality problems. To generate demand in other regions, the US government has published multiple policies and regulations. Some of them are briefly described herein.

1.4.2.1 Federal support for municipal bus electrification

The Federal Transit Administration (FTA) has provision to financially support different transit agencies in purchase of new buses. Federal funds expenditure plan is technology neutral, that is, transit agency is the decision authority on the selection of bus technology. Once the technology is selected, transit agencies can request federal funds for procuring the selected technology. Furthermore, the 116th Congress Assembly has introduced some legislative changes. The Green Bus Act of 2019 (H.R. 2164) has made it mandatory that any bus leased or purchased via FTA funding for public transportation purposes must be a zero-emission bus.

E-buses can be subsidized through various funding programs managed by the FTA. One of such programs is Low or No (Low-No) program. It supports local and state authorities in leasing or purchasing low- and zero-emission transit buses. It also grants funds for acquisition, construction, and leasing of required supporting facilities. Low-No program has a mandatory spending of USD 55 million per year for e-bus purchase through FY2020. Another popular federal funding component is the Bus and Bus Facilities Grant Program (49 U.S.C. 5339). The formula program provides funding for capital expenses, that is, for purchasing buses and for constructing facilities, such as maintenance depots. A new competitive discretionary component was thereafter added to the existing formula program via the Fixing America's Surface Transportation (FAST) Act. Additional funding of USD 300 million per year is now available through the Bus and Bus Facilities Grant Program. In FY2016, the Bus program approved a total of USD 696 million—USD 428 million in formula grants and USD 268 million as discretionary grants. For the FY2020, the total funding was increased to USD 809 million. The formula grant provides each state with a minimum allocation of USD 1.75 million and a minimum of USD 0.5 million are allocated to territories. The remaining funds are disbursed in accordance to the population and service levels.

School buses are not eligible for FTA funding. However, EPA administers a School Bus Rebate program and issues encouragement incentives that assist school districts in replacing older diesel buses with e-buses [18, 19]. In addition, other federal policies such as the Volkswagen Settlement Act can indirectly support vehicle electrification across America.

1.4.2.2 Volkswagen Diesel Emissions Settlement Act 2016

Volkswagen sold thousands of 2.0- and 3.0-L diesel engine vehicles fitted with illegal electronic systems and control devices to manipulate and defeat the emissions tests in the US—violating both the federal and the state laws. The devices were installed in all models starting from the model year 2009. To remedy the harm caused owing to violation of the Federal Clean Air Act by these noncompliant vehicles, Volkswagen Group entered into a series of legal agreements with the US government. The Federal Clean Air

Act requires statewide implementation of plans, which allows the State to accomplish the national 8-h ozone and fine particulate matter standards. Broad deployment of EVA technologies is essential to achieve targets for the 2023–32 period. Therefore, the settlement terms required the Volkswagen Group to

- Pledge USD 2 billion to the Electrify America initiative and build zero-emissions vehicle infrastructure in selected cities. Of the pledged amount, USD 800 million are to be spent in California.
- Fund a USD 2.7 billion national Environmental Mitigation Trust. This fund is available to states affected by the excess emissions caused by the noncompliant vehicles sold by the Volkswagen Group. The states can use this special funding for electrification of their public transportation fleet, including school buses, and for developing the necessary charging infrastructure. California is supposed to receive USD 423 million from this trust.

The investments must be made over a 10-year period, divided in to four 30-months periods [20]. The second 30-month investment period began on July 1, 2019.

1.4.2.3 Policies by California government

State of California has been the most aggressive and a trendsetter in environment reforms in the US. It has pushed the boundaries with its EV policy as well. Californian government has set a target to have 250,000 EV charging stations publicly available by 2025 and to have at least 5 million EVs on the road before the year 2030. Recently, the California Air Resource Board announced that at least 5% of all heavy-duty trucks (Classes 7 and 8) sold in the state by the year 2024 should be electric trucks. The policy mandates 100% transition from diesel to electric trucks should be completed before the year 2045 [21]. Some of the policy reforms that are continually pushing California toward its aim are briefly described here:

The Clean Transportation Program—formerly known as Alternative and Renewable Fuel and Vehicle Technology Program, is a flagship program of the California Energy Commission (CEC). It enables CEC to invest in projects that support state government's climate action goals by advancing adoption and deployment activities of medium- and heavy-duty EVs across the California state. Each year, a new investment plan is drafted by the CEC to determine funding priorities for the upcoming fiscal year. A sum of USD 100 million is then invested every year in projects aiming at the development of sustainable freight system along with clean and efficient technologies capable of reducing GHG emissions and petroleum dependence. The investment portfolio also supports the transition of California's port into "green" ports.

Approximately, half of the particulate matter emissions and 45% of the nitrogen oxide (NOx) emissions in California come from freight vehicles. Emissions from the freight transport system must also be reduced. Challenges for the policymakers in the freight industry are not much different from those confronting the policymakers in the passenger transportation sector. Given these, policymakers and government officials at all levels must establish effective collaboration with stakeholders from industry, labor unions, and other community and environment-related organizations. Strategies must be devised to meet overarching environmental, economic, and mobility goals. Executive Order B-32-15 stresses on the need to transform California's freight transport system and shift to more cleaner modes. Trade Corridors Improvement Fund and Goods Movement Emission Reduction Program (Proposition 1B) were created in the year 2006. Proposition 1B sanctioned USD 1 billion for procuring cleaner freight vehicles and machinery and USD 2 billion for improving the freight transport network in the state. The State committed two dollars to every one sanctioned dollar to the Trade Corridor Improvement Fund. Proposition 1B facilitated the delivery of more than 13,000 clean trucks, locomotives, and marine vessels in the state.

California Sustainable Freight Action Plan was launched in the year 2015 to transform state's freight transport network by accelerating the shift toward zero-emission technologies. Core objectives are

- to improve freight system efficiency by 25%
- to deploy more than 100,000 zero-emission freight vehicle

The plan recognizes state policies and investment designs that can help in achieving the targets.

The Governor of California subsequently released another 10-year investment plan on January 7, 2016. USD 36 billion were allocated to the transportation sector with a 1-year appropriation for clean vehicles—including both passenger and freight transport vehicles and off-road machinery [22].

The California Clean Truck, Bus, and Off-Road Vehicle and Equipment Technology Program: This program is developed and administered by the California Air Resources and the State Energy Resources Conservation and Development Commission. It provides annual funding of USD 12–20 million through December 31, 2020 mainly for development and early deployment of electric trucks (short- and long-haul trucks, vocational trucks), buses, and other off-road machinery. Projects that can be funded include:

- Zero- and near-zero emission truck technology development, demonstration, precommercial pilot, and early commercial implementation.
- Zero- and near-zero emission bus technology development, demonstration, precommercial pilot, and early commercial implementation.
- Purchase incentives for commercially available zero- and near-zero emission truck, bus, and off-road machinery and charging infrastructure.
- Projects that support higher commercial motor vehicle and equipment freight efficiency and larger GHG emissions reductions, including charge management solutions, grid integration technology, and autonomous vehicles.

Retrofitted and remanufactured vehicles that meet emissions and warranty requirements can also receive funding. At least 20% of allocated funds must go toward early commercial deployment of the eligible machinery and vehicles.

1.4.3 India

A large variety of motorized transport modes can be seen plying on the Indian roads. Mobility patterns and auto-segments in India are, therefore, unique and quite different from those common in other parts of the world. Consequently, buses and large goods vehicles, such as trucks, make up only 3% of the total number of on-road vehicles in India. Regardless, more than 70,000 buses of the 430,000 buses, weighing over 6 tonnes, sold globally in the year 2017 were sold in India. In terms of size, the Indian bus market is second only to China [23]. Contribution of Indian bus fleet toward the global emission reduction target is therefore regarded as significant.

To enforce an industry-wide emission reduction directive, a more stringent Bharat Stage (BS) VI emission norm came into effect on April 1, 2020 across India. From this date, no BS IV vehicles, including heavy-duty buses and trucks, will be available for sale in the country. BS emission standards are based on European emission standards for internal combustion vehicles. In-effect, BS VI standard apply much stricter regulation rules to various air pollutants in comparison to the prior standards, namely [24]:

- BS VI allows only 10 parts per million (ppm) of sulfur content as opposed to 50 ppm acceptable limit under BS IV.
- NOx (oxides of nitrogen) emissions are reduced by 25% and 70% for petrol and diesel vehicles, respectively.
- BS VI also requires mandatory integration of on-board diagnostics (OBD) device, that is, a system that provides real-time assessment of powertrain efficiency to the driver, in every vehicle.

Meanwhile, the Government of India is aggressively pushing for electrification of all HDV in India. The National Institute for Transforming India (NITI Aayog), which is the policy think tank of the Government of India and chaired by the Indian Prime Minister himself, aims to achieve 100% electrification of its bus fleet by the year 2030. An independent analytics agency anticipates the number of e-buses in India will cross the 7000 units mark by the year 2025 with a compound annual growth rate of 53% in the period 2018–25 [25]. Together with the national government, various state transport corporations have launched a myriad of policies, such as Smart City Mission, National Electric Mobility Mission Plan (NEMMP), Faster Adoption and Manufacturing of Electric Vehicles in India (FAME), etc. to fund acquisition of e-bus fleets and accelerate their penetration in the city transport network. The text below provides a brief description of these.

1.4.3.1 National Mission for Electric Mobility

National Mission for Electric Mobility is one of the several initiatives taken under a 10-year Automotive Mission Plan (2006–16) by the Department of Heavy Industries, a nodal department of the Government of India for the automotive sector. Its purpose is to fast track adoption and manufacturing of EVs in the country. Mission 2020 document presents a roadmap for replacing 5%–10% of the fleet with EVs, that is, introduce 6–7 million new EVs on road and save 2.2–2.5 million tonnes of fuel by 2020 by ensuring maximum participation of different stakeholders at different stages. A cumulative outlay of INR 14,000 crores is expected during the term of the plan. The roadmap acknowledges different levels of difficulties in implementing the vision for different auto segments. Because introducing e-buses is comparatively difficult than introducing EVs in the two-wheeler segment, critical interventions needed to accelerate adoption are identified, and INR 550 crores is sanctioned as demand incentives for e-buses. Investment of INR 20 crores is approved for building charging terminals for e-buses. Furthermore, a four-phase approach is proposed to support the localization of technology and for developing EV manufacturing capabilities of India. The phases focus on developing R&D capacities, sourcing components locally, developing indigenous products and an EV component ecosystem, creating high capabilities across the value chain, and targeting the export market and foreign investments [26, 27].

1.4.3.2 Faster Adoption and Manufacturing of Electric Vehicles

The Department of Heavy Industries launched phase I of the Faster Adoption and Manufacturing of Electric Vehicles (FAME) scheme on March 13, 2015, with an outlay of INR 795 crores to provide financial incentive and to support the NEMM. The FAME scheme was initially approved for 2 years starting from April 1, 2015. Later, it was extended to cover a period till March 31, 2019, and an additional fund of INR 100 crores was allocated. Benefits were extended to cities that were covered by the "Smart Cities" initiative, major cities of the North Eastern states and union territories. It also supports all other cities with a population larger than 1 million (as per 2011 census), and satellite towns connected to major metro agglomerations, such as Greater Mumbai, Delhi NCR, Hyderabad, and Bengaluru. A total of 465 e-buses were sanctioned to different cities and states in this phase. The demand incentives were based on vehicle size and were disbursed through e-framework created by the department of heavy industries [28].

Phase II of the FAME scheme (FAME-2) started on April 1, 2019. The purpose of this phase of the FAME scheme is to incentivize the purchase of around 7000 e-buses along with 55,000 passenger cars and 1 million two-wheelers. It has been launched for 3 years with a total budget of INR 10,000 crores, of which INR 8596 crores have been earmarked for direct subsidy within this scheme, and the remaining is for building charging infrastructure. Approximately, 10% of the sum was planned to be used in the first year of the phase [29].

Demand incentive for the selected bus is determined from the net present value of all future payments for the bus using a 50% gross capital cost (GCC) model. It is believed that 50% of the GCC rate accounts for the capital cost, and the remaining 50% covers the operational cost. Cost of the bus is estimated by considering the total contract period and the total minimum assured run in kilometers per month. A 10.5% discount rate, compounded monthly, is applied to the cost. The formula used is represented by Eq. (3):

$$\text{Cost}_{\text{Bus}} = \frac{a}{r}\left(1 - \frac{1}{(1+r)^n}\right) \tag{3}$$

where a is the monthly payment for the capital cost and is equal to 0.5*GCC rate*assured monthly kilometer run, r is the monthly discount rate, and n is the contract period in months. A subsidy equivalent to 40% of the cost estimated by Eq. (3) is available as demand incentive for a bus. In any case, the maximum demand incentive available for buses of different lengths under the FAME-2 program is specified further:

Standard bus (length between 10 and 12 m)	INR 55,00,000
Midibus (length between 8 and 10 m)	INR 45,00,000
Minibus (length between 6 and 8 m)	INR 35,00,000

Only the OEMs with manufacturing facilities in India can avail this subsidy. Furthermore, the OEMs are required to have completed testing and certification of at least one model of the standard/midi/mini e-bus in accordance to the Central Motor Vehicle Rules 1989 from a designated testing center in India to be eligible for the subsidy under FAME-2 program. The total incentive is released in three installments (20-40-40). The first installment (of 20%) becomes due on successful issue of supply order and signing of an agreement between the state transport unit and the selected bidder; the second (40%) upon delivery of buses; and the final installment is released 6 months after commercial deployment of the buses.

To begin with, deployment of 5000 e-buses was initially planned. The Ministry of Heavy Industries and Public Enterprise invited an expression of interest from different states and cities. The deployment plan is given in Table 7 [30].

However, in August 2019, the Ministry authorized the purchase of 5595 e-buses in 64 cities under the FAME-2 scheme, that is, more than planned, for intracity and intercity operations [31].

With their support, Himachal Pradesh became the first Indian state to acquire a fleet of e-buses. Twenty-five 9-m buses were procured jointly by the Central government and the State government, with a 75–25 stake division, to operate on a 51-km stretch between Manali and Rohtang region. Each bus costed around INR 17 million [32].

Table 7 E-bus deployment plan of the Indian government under the FAME-2 scheme.

City category	Targeted cities	Cities to be selected	Min. buses to be sanctioned to each city	Total buses sanctioned
Population > 4 million	8	5	300	1500
Population > 1 million	45	20	100	2000
Special category of states	25	10	50	500
Other cities	50	20	50	1000
Total		**55**		**5000**

Interesting to note is that the buses smaller than 10 m seem to dominate the Indian e-bus market, and accounted for more than 60% of the total sales in the year 2018. This is mainly owing to the length agnostic subsidy offered in the country by various State Road Transport Undertakings. Other cities are not too far behind. For example, Bengaluru city has procured 1500 e-buses through public-private partnership. In March 2019, Delhi Transport Corporation procured 1000 e-buses for its public transport fleet. In addition, the Delhi Integrated Multimodal Transit System is overseeing construction of 1000 e-bus depots in the state. To reduce their carbon footprint, the Brihan Electric Supply and Transport undertaking (BEST) in Mumbai added ten 9-m long e-buses to its fleet in September 2018. Subsequently, a request for proposal was issued through FAME-2 channel to supply and operate 340 e-buses (140 standard buses and 200 midibuses) on a gross cost contract model. Responsibility of route maintenance lies with BEST. Similarly, the Telangana State Road Transport Corporation (TSRTC) welcomed 40 e-buses in its fleet and is planning to commission another 334 12-m long, low floor e-buses to operate in the Greater Hyderabad zone and in Warrangal area under the Smart Cities Mission. Pune city has also an e-bus program coordinated by the Pune Mahanagar Parivahan Mahamandal Limited (PMPML). It acquired 150 e-buses through the FAME-2 scheme. The city has further issued work orders to acquire 500 more such buses sponsored by Pune Smart City Development Corporation Limited [33].

1.4.3.3 Make in India

Considering the EV proliferation rate in India and the market readiness, the Government of India is incentivizing those who want to foray into manufacturing of EVs and EV components. The goal is to create an environment favorable for the transition from internal combustion engine vehicles to EVs and to create employment opportunities on both sides of the EV market, that is, supply side and demand side by attracting investments in EV manufacturing domain and by limiting the import of ready-to-sell EVs. The government is emphasizing on at least 40% localization to create a domestic supply chain. Therefore, special incentives and concessional packages, including land, power, and water needed for battery manufacturing plants, have been announced by various states.

Import duties in India are levied according to the value added to the finalized product locally in India, that is—higher the value addition made by the Indian companies, lower the customs duty. Accordingly, imported finished goods that are already in a ready-to-sell form and require the lowest value addition from the Indian business ecosystem attract the highest custom duty. In contrast, raw materials and components that require a comparatively higher value addition before they can be sold to the end-user are applied the lowest, in some cases, amounting to zero customs duty. In line with this, battery cells, motors, controllers, magnets, connectors, and integrated circuits have the lowest custom duty. Furthermore, subsystems, such as battery packs, power modules, drive train, and air conditioners, have a higher custom duty, and finished electric cars have the highest custom duty. For battery packs, battery chemicals account for 35%–40% of the total value of the battery pack. Cell manufacturing contributes 25%–30% of the value addition, and battery pack manufacturing or assembly of packs from cells adds 30%–40% to the final value. Furthermore, goods and service tax (GST) policy in the country promotes higher vehicle utilization.

Analysts believe that India needs 50 GWh of battery cells by 2025. Therefore, the Indian government is offering short-term and long-term tax incentives and faster depreciation as incentives to energy operators, that is, businesses providing battery charging and swapping service, to deploy standard and fast chargers across the country and to create battery swapping centers. GST for the chargers is set at the same level as that of the vehicle. Moreover, swappable batteries and bodies of EVs that accept swappable batteries are treated similarly for GST purposes in the current policy [34].

1.4.4 Europe

The concept of "eMobility" is not new to the public transportation system in Europe. Electrified modes of transport, such as trams, trolleybuses, and metros, have existed in cities such as Geneva and Moscow for several years now. Other cities are also finding synergy and integrating e-buses in their existing transport network at a decent pace. There were 1273 operational e-buses in the European region in mid-2018. Since then, the uptake of e-buses by the European market is constantly on the rise [35]. The European countries procure 20,000 new buses every year, 10% of which are e-buses. Forecasts indicate that the e-bus market share will increase from 10% to 12% presently to approximately 40% in the year 2025. However, the majority of the e-bus sales in Europe is concentrated in the following six countries—the United Kingdom (UK), the Russian Federation, the Netherlands, Germany, Poland, and France. Together, these six nations account for around 56% of all orders in Europe [36]. The primary reason is strong support and forward-thinking policies of the local government.

1.4.4.1 United Kingdom

The UK government provides funding for purchasing low- and zero-emission vehicles through a range of grants. A few examples are as follows:

The Bus Service Operators Grant Low Carbon Emission Bus Incentive program was launched in the year 2009. Bus operators are paid 6 pence per kilometer to cover the high operational cost of low carbon emission buses.

The Green Bus Fund assisted the Department of Transport in the deployment of 1250 low-emission buses in England between the years 2009 and 2013.

Low Emissions Bus Scheme: the success of the Green Bus Fund program encouraged the Department of Transport to award GBP 30 million to promote uptake of e-buses by 13 local authorities and operators in the period 2016–19. A total of 300 e-buses have been added to the fleets in England and in Wales. Additional GBP 49 million have been reserved for procurement of more buses in the next funding round. Moreover, GBP 40 million have been allocated to the Clean Bus Technology Fund.

1.4.4.2 The Netherlands

By the end of the year 2018, there were approximately 3200 e-buses running on roads of different cities of Netherlands as a part of its public transport network. In addition, there were more than 400 commercial vehicles, for example, lorries on-road. Netherlands is working hard to extend its leadership position in electric transport further and is exporting Dutch-made heavy-duty EVs, for example, electric lorries and buses globally. In 2016, the Ministry of Infrastructure and Water Management signed an administrative agreement with the 14 public transport authorities in the Netherlands, which mandates all the new buses entering service from the year 2025 onward to be electric and to replace their entire fleet with e-buses by the year 2030. Moreover, all energy needed for charging these e-buses must be generated regionally from renewable sources using solar panels and wind turbines.

To stimulate sustainable transport development, the Dutch central government and local authorities started the Zero-Emission City Logistics Green Deal in the year 2014. It allows car manufacturers, shipping companies, and other stakeholder organizations to enter into an agreement with the local government under private law. "Green Deal" objectives are divided into nine themes; energy and mobility are two of them. There are over 100 participants in this initiative, now all sharing the same goal of establishing optimum zero-emission city logistics by the year 2025. As a part of this deal, significant steps have been taken to create zero-emission zones in major cities such as Hague, Rotterdam, and Utrecht. These include deployment of zero-emissions cleaning vehicles across the country. The Netherlands has a fleet of 6000 cleaning vehicles, comprising large lorries used for domestic waste collection and small street cleaners used for cleaning the inner-city region. Within the inner-city region, freight movement increases annually by 2%–5% owing to changing customer requirements and digitization. Increasing HDV movement affects air quality and safety levels in the city. The Directorate General for Public Works and Water Management has signed a letter of intent to replace all the freight vehicles and the cleaning vehicles with EVs, charged with renewable energy, by the year 2030. To facilitate this, a subsidy of 40,000 € on purchase price of an e-truck is available in Amsterdam region [37].

In the Netherlands, it is not only the central government that is tasked with planning and finding optimum solutions. Much success is derived from partnerships—consultations with private parties lead to creation of well-considered generic policies. The key factors contributing to the success of the Dutch model are cooperation, knowledge development, and regional anchoring.

1.4.4.3 Poland

Polish e-bus market is one of the largest e-bus markets in Europe in terms of operating and contracted e-buses. Local bus operators in all cities with a population larger than 50,000 are required to have a minimum of 5% share of e-buses in the composition of their fleets before the end of the year 2021. Poland is also among the fastest growing markets owing to the support it gets from various Ministries and government institutions.

The Ministry of Entrepreneurship and Technology launched an e-bus design, build, and deployment program: Polish Electric Bus in the year 2017. The primary requirement was to build the bus using indigenous components, that is, to have the key components of the bus developed locally in Polish R&D centers and manufactured locally by Polish suppliers. Strategic goal was to create a local demand and supply chain, an e-bus market worth €583 million by the year 2025. The program coordinator was the Polish Development Fund. It has now created a functional ecosystem through multiple cross-sectoral cooperation schemes and boasts of a predominantly well-developed supply market—not only of e-buses but also jobs. The OEMs and partner companies involved in the creation of 5000 new jobs in addition to 1000 new e-buses annually. In the coming years, the national electromobility plan is expected to contribute 1.1% to the GDP growth and create at least 81,000 new jobs every year. Therefore, the Polish government has declared financial incentives worth €2.3 billion to encourage uptake of electric public transport in the period 2018–28. The Poland Low Emissions Transport Fund supports it. The introduced legislations provide real financial mechanisms and instruments to accelerate eMobility development. Measurable effects can be produced in the construction and expansion of charging infrastructure through funding available at the level of up to 50% of the eligible costs. With regards to purchase of an e-bus, the grant will cover 55% of the eligible costs. Maximum financial support is limited to PLN 1,045,000 per bus. Besides, the direct subsidy can be up to PLN 36,000 only. Other sources of public financing for building up e-bus fleets include Emissions Free Public Transport Program and European Structural and Investment Funds [36].

The number of heavy-duty EVs is continuously on the rise in the EU region. This growth is supported by a report from the EU Commission, which states that 25% of the CO_2 emissions from the road transport come from buses and trucks [38]. The European Council has, thus, adopted binding targets for all Member States in the award of public contracts. For example, Germany is obliged to procure at least 10% and 45% emissions-free heavy commercial vehicles and buses, respectively. From the year 2026 onward, the minimum binding percents are 15 and 65 for the two respective segments. More details are available under the "Clean Vehicles Directive" [39].

Similar initiatives are driving the growth of heavy-duty EV sector in South American countries such as Argentina and Chile as well [40].

1.5 Summary

Electrification of public and freight transport systems is crucial for meeting the climate action goals. This chapter provides an overview of various standards and regulations established by different governing bodies to guide the design and development process of EVs. They are also used to measure the delivered performance and to assess their safety ratings. In addition, a comprehensive discussion on a variety of subsidies announced by different countries and agencies to accelerate the introduction and adoption rate of heavy-duty EVs around the world is offered. Noteworthy is that these subsidies and different regulations are connected to vehicle classification according to weight. To that end, the foremost challenge while deploying heavy-duty EVs is to maintain the same level of service as an equivalent diesel bus or truck. A practical example of this challenge is the passenger cabin design in e-buses, that is, with a fixed GVWR despite the additional weight of the battery packs, how can we maintain equivalent passenger capacity and comfort in e-buses?

Finding the right technology and solution for the targeted application is the key to the successful integration of e-buses and e-trucks in existing public transportation networks, that is, without any negative impact on their daily performance. The remainder of this book provides information that serves this purpose.

References

[1] United Nations, Global Technical Regulation No. 20, Electric Vehicle Safety (EVS) (ECE/TRANS/180/Add.20), United Nations, 2018. GE.18-07039(E).

[2] European Parliament, CO_2 Emissions From Cars: Facts and Figures, 2019, (cited 2020 June 20). Available from: https://www.europarl.europa.eu/news/en/headlines/society/20190313STO31218/co2-emissions-from-cars-facts-and-figures-infographics.

[3] A. Islam, N. Lownes, When to go electric? A parallel bus fleet replacement study, Transp. Res. Part D: Transp. Environ. 72 (2019) 299–311.

[4] M. Hughes-Cromwick, M. Dickens, 2020 Public Transportation Fact Book, seventy-first, American Public Transportation Association, 2020.

[5] M. Naumanen, T. Uusitalo, E. Huttunen-Saarivirta, R. van der Have, Development strategies for heavy duty electric battery vehicles: comparison between China, Eu, Japan and USA, Resour. Conserv. Recycl. 151 (2019) 104413.

[6] S. Arora, A. Kapoor, Mechanical design and packaging of battery packs for electric vehicles, in: G. Pistoia, B. Liaw (Eds.), Behaviour of Lithium-Ion Batteries in Electric Vehicles: Battery Health, Performance, Safety, and Cost, Springer International Publishing, Cham, 2018, pp. 175–200.

[7] EPA, Vehicle Weight Classifications for the Emission Standards Reference Guide, Emission Standards Reference Guide, United States Environmental Protection Agency, 2017.

[8] M. Murray, Commercial Motor Vehicle Classification—Gross Vehicle Weight Ratings for Trucks, Supply Chain Management—Small Business 2019 (cited 2020 June 12). Available from: https://www.thebalancesmb.com/commercial-motor-vehicle-classification-2221025.

[9] J. Du, M. Ouyang, X. Wu, X. Meng, J. Li, F. Li, Z. Song, Technological direction prediction for battery electric bus under influence of China's new subsidy scheme, J. Clean. Prod. 222 (2019) 267–279.

[10] S. Shengyang, Trends and Challenges in Electric-Bus Development in China, 2018, Electromobility—Urban Transport (cited 2020 July 3). Available from: https://www.sustainabletransport.org/archives/5770.

[11] FCEB, Codes/Standards/Regulations, 2019, (cited 2020 June 10). Available from: https://www.fuelcellbuses.eu/wiki/codesstandardsregulations-framework/codesstandardsregulations.

[12] V. Ruiz, A. Pfrang, A. Kriston, N. Omar, P. Van den Bossche, L. Boon-Brett, A review of international abuse testing standards and regulations for lithium ion batteries in electric and hybrid electric vehicles, Renew. Sust. Energ. Rev. 81 (2018) 1427–1452.

[13] H. Zhang, X. Song, T. Xia, M. Yuan, Z. Fan, R. Shibasaki, Y. Liang, Battery electric vehicles in japan: human mobile behavior based adoption potential analysis and policy target response, Appl. Energy 220 (2018) 527–535.

[14] Ministries of China, Measures for the Promotion and Application of New Energy Buses (Trial), 2015, Policies and Regulations (cited 2020 July 4). Available from: http://www.mof.gov.cn/mofhome/yunnan/lanmudaohang/zhengcefagui/201609/t20160909_2413911.html.

[15] Z. Song, Y. Liu, H. Gao, S. Li, The underlying reasons behind the development of public electric buses in China: the Beijing case, Sustainability 12 (2) (2020) 688.

[16] Department of Transportation Services China, Notice of the Ministry of Transport on Printing and Distributing the "Thirteenth Five-Year Plan for Urban Public Transport Development", 2016, (cited 2020 July 3). Available from: http://www.mot.gov.cn/zhuanti/shisanwujtysfzgh/guihuawenjian/201702/t20170213_2163887.html.

[17] J. Du, F. Li, J. Li, X. Wu, Z. Song, Y. Zou, M. Ouyang, Evaluating the technological evolution of battery electric buses: China as a case, Energy 176 (2019) 309–319.

[18] B. Canis, C.E. Clark, M.F. Sherlock, Vehicle electrification: federal and state issues affecting deployment, Congressional Research Service (CRS Report), 2019.
[19] W. Mallett, Federal Public Transportation Program: In Brief, Library of Congress, Congressional Research Service, 2020, p. R42706.
[20] B. Canis, A. Vann, R.K. Lattanzio, B.D. Yacobucci, Volkswagen, Defeat Devices, and the Clean Air Act: Frequently Asked Questions, Congressional Research Service, 2016.
[21] E. Miller, Carb Updates Proposal to Speed California's Transition to Zero-Emission Vehicles, 2020, Transport Topics (cited 2020 June 15). Available from: https://www.ttnews.com/articles/carb-updates-proposal-speed-californias-transition-zero-emission-vehicles.
[22] E. Brown Jr., California Sustainable Freight Action Plan, Office of the Governor of California, 2016. July www.casustainablefreight.org.
[23] R. Scriven, Indian electric bus market, what's going on? News (2018). (cited 2020 June 20). Available from: https://www.sustainable-bus.com/news/indian-electric-bus-market-whats-going-on/.
[24] Divya, Bharat Stage 6 (BS6) Emission Standards: Everything You Need to Know, JCBL Limited, 2019. (cited 2020 June 25). Available from: http://www.jcbl.com/blog/bs6-emission-standards.aspx.
[25] L. Wood, India Electric Bus Market Size, Share, Development, Growth, and Demand During the Forecast Period, 2019-2025, 2019, Globe Newswire—Research and Markets (cited 2020 June 25). Available from: https://www.globenewswire.com/news-release/2019/12/30/1964871/0/en/India-Electric-Bus-Market-Size-Share-Development-Growth-and-Demand-During-the-Forecast-Period-2019-2025.html.
[26] IEA, National Electric Mobility Mission Plan (Nemmp), 2019, Policies (cited 2020 June 20). Available from: https://www.iea.org/policies/6201-national-electric-mobility-mission-plan-nemmp.
[27] DHI, The National Electric Mobility Mission Plan 2020, 2017, Automotive Industry (cited 2020 June 25). Available from: https://dhi.nic.in/UserView/index?mid=1347.
[28] PressInformation Bureau, Fame India Scheme, 2019, Ministry of Heavy Industries & Public Enterprises, Government of India (cited 2020 June 25). Available from: https://pib.gov.in/newsite/PrintRelease.aspx?relid=191377.
[29] IEA, Faster Adoption and Manufacturing of Hybrid and Ev (Fame) II, 2019, Road Transport—Policies (cited 2020 June 25). Available from: https://www.iea.org/policies/7450-faster-adoption-and-manufacturing-of-hybrid-and-ev-fame-ii.
[30] DHI, Expression of Interest Invitingproposals for Availing Incentives Under Fame India Scheme Phase II for Deployment of Electric Buses on Operational Cost Model Basis, Ministryof Heavy Industries & Public Enterprises, Governmentof India, F.No.6(09)/2019-NAB.II(Auto), 2019.
[31] S. Shanthi, Paving the Way for Emobility: State and Central Government Ev Policies in India, 2019, Inc 42—Features (cited 2020 June 25). Available from: https://inc42.com/features/paving-the-way-for-emobility-state-and-central-government-ev-policies-in-india/.
[32] UITP, India Launched Its First Electric Buses Service, 2017, India (cited 2020 June 25). Available from: https://india.uitp.org/articles/india-launched-electric-bus-service.
[33] S. Pothula, From Delhi to Pune, How Indian Cities Are Going Green by Adopting Electric Buses, 2020, Social Story—Transportation (cited 2020 June 25). Available from: https://yourstory.com/socialstory/2020/03/delhi-mumbai-pune-electric-buses-climate-change.
[34] S. Juyal, H. Sanjeevi, A. Saxena, S. Sharma, A. Singh, S. Chander, A. Jhunjhunwala, Zero Emission Vehicles (Zevs): Towards a Policy Framework, 2018. MOVE—Global Mobility Summit.

[35] L. Mathieu, Electric Buses Arrive on Time. Marketplace, Economic, Technology, Environmental and Policy Perspectives for Fully Electric Buses in the EU, Transport and Environment, 2018.
[36] D.P. Patella, A. Perchel, I. Jaques Goldenberg, J. Lee Brown, M. Baker, O.J.E. Joy, C. Amato, R.O. Steinmetz, R. Van der Ploeg, E. Breen, Z. Koks, B. Aritua, Y. Yang, H. Deng, E.A. Beukes, L. Qu, A. Hoyos Guerrero, P. Turner, H. Fang, A. Damasceno Ferreira Junior, Electric Mobility & Development: An Engagement Paper from the World Bank and the International Association of Public Transport, World Bank Group, Washington, D.C., 2018. http://documents.worldbank.org/curated/en/193791543856434540/Electric-Mobility-and-Development-An-Engagement-Paper-from-the-World-Bank-and-the-International-Association-of-Public-Transport.
[37] rvo.nl, Mission Zero—Powered by Holland, Netherlands Enterprise Agency, The Ministry of Economic Affairs and Climate Policy, 2019. (cited 2020 June 12). Available from: www.rvo.nl.
[38] C. Hampel, Eu states set binding Co2 limits for trucks and buses, Politics (2019). Available from: https://www.electrive.com/2019/06/15/eu-states-set-binding-co2-limits-for-trucks-and-buses/.
[39] European Commission, Clean vehicles directive, in: Mobility and Transport, 2020. Clean Transport, Urban Transport (cited 2020 July 1). Available from: https://ec.europa.eu/transport/themes/urban/vehicles/directive_en.
[40] C. Randall, Zebra to Bring Electric Buses to South America, 2020, (cited 2020 July 4). Available from: https://www.electrive.com/2020/04/07/zebra-to-bring-electric-buses-to-south-america/.

PART I

Drivetrain

CHAPTER 2

Drivetrain Configurations for Heavy-duty Electric Vehicles

2.1 Introduction

The idea of developing electric vehicles (EVs) instead of fossil fuel vehicles goes back to the 18th century. Since then, scientists and manufacturers have attempted many times to design EVs. Robert Anderson built the first crude electric carriage in 1839, and David Salomon built the first electric car in 1870. Edison e-bus in 1915 was the first EV used for the public transportation [1]. Heavy batteries and poor performance of electric motors were the main limitations of the first-generation EVs. Later on, interest on EVs diminished owing to the development of electric self-starters for the gasoline vehicles and low price of oil, until early 1980s when environmental and GHG emission concerns raised [2].

Hybrid electric vehicles (HEVs) are more popular compared to pure EVs these days owing to better mileage and lack of publicly available battery charging infrastructure. Many countries in Europe and Asia have defined new standards and regulations for efficiency and carbon emissions of motor vehicles to reduce the GHGs in the transport sector. Therefore automobile manufacturers are actively involved in research and development of EVs (passenger cars, city, and intercity buses) for future market in the past decade.

Conventional EVs have a central electric motor that drives two or all four wheels of the vehicle through a reduction gear or directly [3]. Using two or four electric motors to drive 2/4 wheels independently has been a popular approach recently, especially in high-performance vehicles such as electric racecars. In-wheel motors (motor is embedded inside the tire rim) are also of interest for high-performance EVs in the recent years. Wellington Adams introduced the in-wheel motors first in 1884. He built and attached an electric motor directly in the wheel through complicated gearings [4]. Individual torque control of each wheel is made possible by

Heavy-duty Electric Vehicles
https://doi.org/10.1016/B978-0-12-818126-3.00003-8

using in-wheel motors or hub motor. Therefore better vehicle speed/torque and acceleration control can be achieved that consequently improves electric drivetrain's dynamic stability, efficiency, and safety [5].

Different electric motors such as brushed direct current (DC) motors, induction motors (IM), switched reluctance motors (SRM), and permanent magnet synchronous motors (PMSM) have been used as the propulsion system in EVs for different applications. Some of the requirements of the suitable electric motors for EV applications are shown below [6]:

- High torque at low speeds
- High torque/power to size ratio
- Constant power in a wide speed range
- High efficiency
- High dynamic response
- High controllability
- High robustness and reliability
- Low electromagnetic interface (EMI) noise generation
- Reasonable cost

PMSMs are more prevalent in high-performance EVs owing to their high efficiency and faster dynamic torque response; however, they have a complex control system and higher manufacturing cost owing to permanent magnet price in the market [3]. Detailed technical discussions on electric motors and their comparison are beyond the scope of this chapter; however, if the reader is interested, then please refer to references [3, 4] for more discussions.

EV drivetrain control capabilities, efficiency, and safety factors are more critical in HDVs such as e-buses because they have a high inertia at full load (in some cases, up to 100 passengers onboard) and consequently require a higher amount of battery energy onboard when compared with passenger cars. Moreover, they are required to comply with safety requirements, maintain speed under full load condition on 5%–12% ramp, and avoid hazards to the environment. Vehicles meant for interstate operations are designed to run on multilane highways with a maximum grade of 3%. Vehicles operating on off-highways can encounter intermittent grades of up to 12%. During off-highway operation, 8% grade is relatively common [7]. In this chapter, heavy-duty EV drivetrain requirements are discussed and compared for various available EV drivetrain configurations in the market.

2.2 Heavy-duty electric vehicle drivetrain

An EV drivetrain is defined as a group of subsystems that is used to convert the electrical energy stored in the battery to the mechanical torque that is delivered to the vehicle's wheels in different configurations. These subsystems include electric motors, motor controllers/inverters, reduction gearbox, and differential (if applicable).

Consideration of performance requirements is the first point to define the vehicle propulsion system specifications. In general, EV drivetrain must be capable of frequent starts, stops, and reverse rotation, with maximum output torque. In addition to that, a heavy-duty EV should be able to start from halt position and repeatedly accelerate smoothly in a short time to overcome the high inertia of the load on various road inclinations [8].

The other major factor is the efficiency that directly affects the mileage of the vehicle, the amount of onboard battery energy, size and weight, and ultimately, the vehicle cost. Although electric motors have higher efficiency when compared with diesel engines, the torque produced in the motor will be delivered to the wheels through multiple torque converting stages such as gearbox and differential. Therefore the overall EV drivetrain efficiency highly depends on the drive cycle profile; for example, the average operational efficiency of a torque converter in vehicles during city traffic conditions is less than 60% [9]. Manhattan Bus Cycle is a good platform for modeling analysis in this regard. It was developed for urban buses based on actual observed driving patterns of urban transit buses in the Manhattan, New York City. This cycle consists of frequent stops and very low speed of over 1089 s, with the maximum speed of 40.88 km/h.

2.2.1 EV drivetrain configurations

Heavy-duty EVs drivetrain could be divided into five main configurations/ concepts as discussed and shown further.

Concept 1: Central motor + single ratio gearbox + differential

This concept is the most common configuration used in e-buses and trucks so far. As shown in Fig. 1, a single motor generates the required torque that is then converted through a fixed-ratio gearbox and a differential. Rated speed of the selected motor should match the maximum required vehicle speed according to the final reductions gears. Solid lines show HV lines, and dashed lines represent the communication area network (CAN) bus layout.

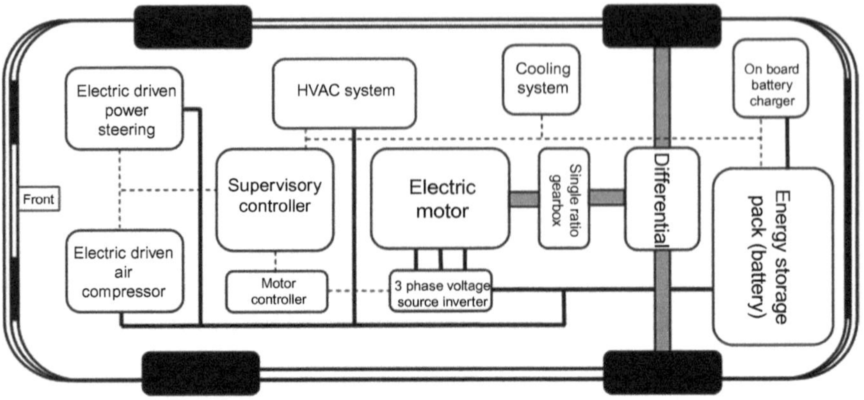

Fig. 1 Central motor + single ratio reduction gearbox + differential.

Concept 2: Central motor + multiple ratio gearbox + differential

As shown in Fig. 2, this concept is similar to the previous configuration except for the gearbox gear ratios. This concept is also common because using multiple gear ratios reduces the maximum required rated torque of the selected electric motor because the gear ratio is higher at low speeds. However, it adds the complexity of the automatic gearbox control to the system.

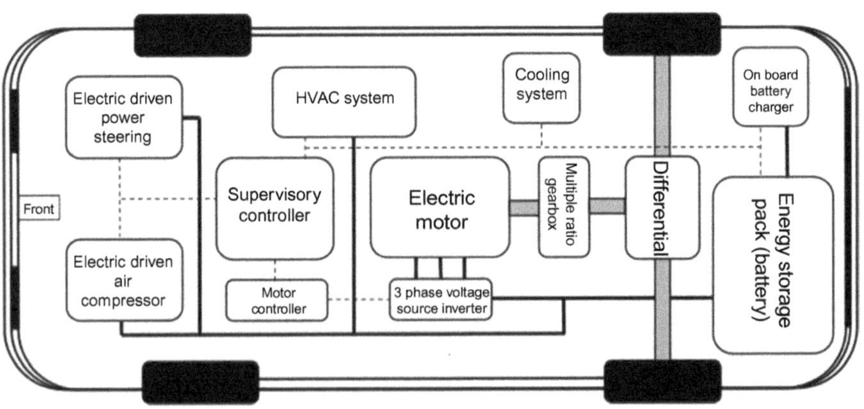

Fig. 2 Central motor + multiple ratio gearbox + differential.

Concept 3: Central motor + differential

As shown in Fig. 3, there is no reduction gearbox in this concept. Therefore the selected electric motor must be able to generate high torque at low speeds, while it does not need to have a high speed.

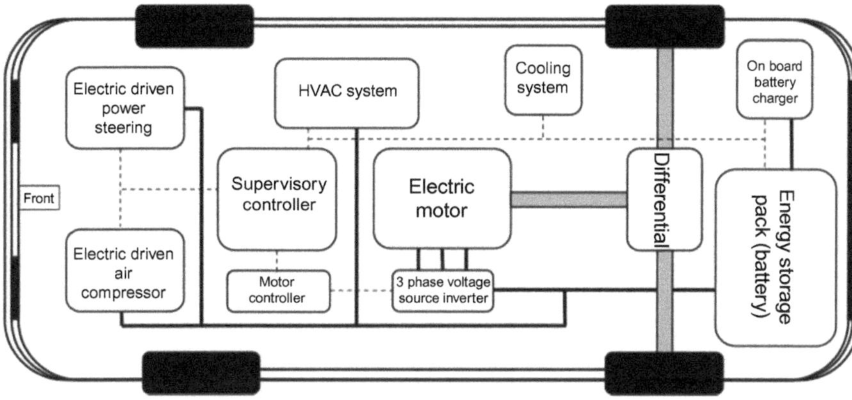

Fig. 3 Central motor + differential.

Concept 4: Two by-wheel or hub motors + single ratio gearbox

In this concept, two motors drive rear wheels separately through a reduction gearbox as shown in Fig. 4. Therefore the selected electric motors power rating is lower than the previous concepts. This configuration is of interest for HDV axle manufacturers within the last decade, and there are multiple off-the-shelf solutions for e-buses or trucks from various manufacturers in the market. However, it has a more complex control system because both motors need to be synchronized accordingly.

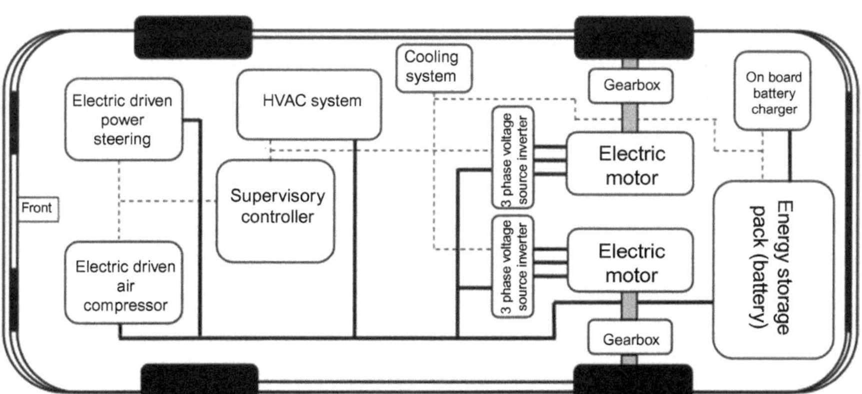

Fig. 4 Two by-wheel or hub motors + single ratio gearbox.

Concept 5: Two in-wheel motors

As shown in Fig. 5, in-wheel motors are directly located inside the wheel. This feature is not common in HDV owing to high torque requirements and added weight of the motors to the wheels. However, some companies have developed such in-wheel motors for passenger cars.

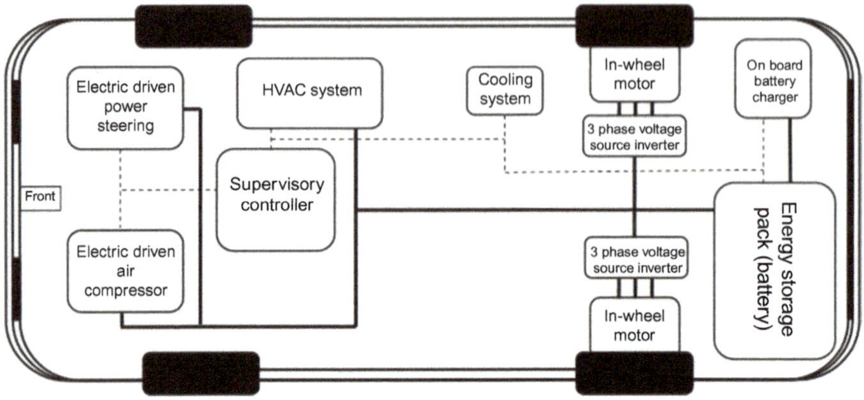

Fig. 5 Two in-wheel motors.

2.2.2 EV drivetrain requirements

EV drivetrain specifications are highly related to the type of vehicle application and performance requirements. For example, the performance requirements of an electric racecar are different from an e-bus. The racecar requires high-power density battery and a fast torque response motor and controller that can deliver the peak torque in wide speed range to fulfill the high acceleration and deceleration performances required. Torque commands frequency in race applications is usually between 100 and 200 Hz. In comparison, an e-bus needs a high-energy density battery pack for the better mileage, plus an electric motor with high torque at low speeds to overcome the high inertia of the vehicle body during frequent starts and stops driving scenarios. Torque commands frequency in e-bus applications is generally between 3 and 10 Hz. However, some factors are commonly required for all the applications such as efficiency, high power to size/weight ratios, and low manufacturing cost.

In addition to the parameters above, there are other critical selection criteria related to the mechanical build and assembly of the drivetrain in the vehicle chassis. Items given below are a summary of the most vital factors that are important for the selection of an EV drivetrain for a heavy-duty design and development project.

- **Powertrain efficiency**: This includes efficiency of all subsystems that are involved to convert the electrical power to mechanical power and its transfer to the wheels: motor controllers (inverters), electric motors, gearbox, and differential. In general, a smaller number of torque converters results in higher efficiency.

- **Modularity**: It is the degree to which the EV drivetrain subsystems and components are independent and interchangeable for future upgrades or any modifications as required. This is of interest for EV developers.
- **Drivetrain control**: The ideal scenario is to control the overall EV drivetrain performance as simple as possible to reduce the control system development complexity. In general, adding electronic control subsystems increases the overall drivetrain control structure.
- **Retrofitting simplicity**: It is the required mechanical work to assemble the EV drivetrain on the current vehicle chassis. This also includes required vehicle chassis and body modifications.
- **Ease of servicing**: The primary considerations are simplicity and accessibility of EV drivetrain subsystems for maintenance or potential change owing to malfunctioning.
- **Low mass and volume**: Smaller and lighter components are much easier to be accommodated and assembled owing to a limited and confined area in the vehicle structure.
- **Low vibration**: Lowest mechanical vibration is the ideal situation for the drivetrain assembly and performance.
- **Low acoustic noise**: High voltage and high-power motors and inverters are the primary sources of acoustic noise, while gearbox and differential add to this. Mechanical noise, aerodynamic noise, and electromagnetic noise produced by the air gap magnetic flux waves are the main noise sources in an electric motor [10]. In general, electric HDVs are more quiet when compared to the equivalent diesel vehicles.
- **Low drivetrain cost, market availability and lead-time**: Same as all other engineering projects, cost is one of the main factors. Also, the design and development phase is limited, and the components with the lowest manufacturing lead-time are more desired. While the EV drivetrain solution needs to have a reasonable cost and lead-time, the components and subsystems must be available/accessible in the market for future serial production.
- **Low servicing time and cost**: Following the pervious item, maintenance services also need to be done at a reasonable cost and time. In general, the servicing cost of an electric city bus is much lower than a diesel bus. Some published reports claim that e-buses could save up to $400,000 in fuels consumption and up to $125,000 in services in the lifetime [11]. Maintenance cost comparison studies

between manufactured e-buses and conventional diesel buses estimate \$563/month saving for BYD bus, \$938/month saving for Proterra bus, and \$861/month saving for Xcelsior bus [12].

- **Robustness, fault handling, and durability**: This is a critical item in EV applications owing to the existence of HV and safety regulations around that. Practically, the overall EV driveline system needs to be designed in such a way as to achieve a fail-safe mode operation.
- **Compliance to EV standards**: Following the previous item, all EV drivetrain subsystems must comply with the existing IEC and ISO standards related to EV system design and development.

2.3 EV drivetrain comparison

In this section, we are going to compare the five heavy-duty EV drivetrain concepts presented earlier according to the selection criteria that were discussed in the previous section. For rigorous comparison, numeric values are assigned to each concept for every selection criterion according to the comparison discussion as further.

1: Very bad/low (not suitable)
2: Bad/low (not ideal)
3: Neutral/normal
4: Good/high (suitable)
5: Very good/very high (ideal)

The selection criteria are also categorized into the highest, moderate, and least priority items based on practical design and development steps for heavy-duty EVs. Then, each category has a factor as below to emphasize on the importance of the selection criteria. The values are multiplied by the category priority factor and added to show which EV configuration concept gains the highest points. The comparison discussions are based on the practical experience of the authors on multiple design and development of e-buses and trucks.

- Highest priority category (Factor 3)
  - Efficiency
  - Modularity
  - Low mass/volume (ease of packing)
  - Low drivetrain cost
  - Low servicing time and cost
  - Control simplicity

- Moderate priority (Factor 2)
  - Retrofitting simplicity (time/mechanical)
  - Ease of servicing (accessibility/complexity)
  - Robustness and fault handling (durability)
  - Compliances/electromagnetic interference (EMI)
- Least priority (Factor 1)
  - Low acoustic noise
  - Low vibration

High number of power conversion steps reduces the efficiency of the overall EV drivetrain. Therefore, a single central motor that is directly connected to the differential has a higher efficiency when compared to concepts 1 and 2 that have an additional gearbox. In-wheel motors have a higher efficiency because the motor is embedded inside the rim and have minimum power conversion steps.

From modularity point of view, concepts 1, 2, and 4 are the best because they need a motor with higher speed and lower torques that are popular in the EV market. On the other hand, concept 3 needs a high torque, low-speed motor that is manufactured by limited companies. In-wheel motors are also highly customized and limit the upgrade option of the system in future.

Less number of components means lower volume and weight. In general, high-power motors have higher weight and volume, so concept 3 motor is much bigger than concepts 1 and 2, but it does not need a gearbox. Concept 4 has lower power motors but two sets plus the gearbox. In-wheel motors are the most compact solutions; however, their weight is still similar to the concept 4 motors.

From the cost point of view, in-wheel technology has the most expensive motors. In general, higher power motors are more expensive. Also inclusion of more components will add to the overall cost of the system. The servicing time and cost are related to the number of components. However, the mechanical system maintenance requirement is much more than electrical motors. Any general service requirement that involves tire affects in-wheel motors that are not good.

From the drivetrain control complexity, in-wheel motors are the most complex owing to high dynamic torque response in PMSMs that is directly connected to the wheel. Using multiratio gearbox and multiple electric motors adds to the control complexity as well. Therefore concepts 2 and 5 have the most complex control system structure.

Retrofitting simplicity highly depends on the number of components and the type of available vehicle chassis. In concept 5, installation to the chassis is complex owing to the specific requirements of in-wheel motors. In general, the correct alignment of the reduction gear and the differential is a complex task that needs to be done. Concept 4 needs more chassis modification owing to the two motors and gearbox installation. EV configuration that has a more complicated installation process has more complex maintenance process as well.

All HV electrical components such as motor and motor controller may fail owing to various electrical and mechanical faults. Some faults cause the subsystem failure but do not damage the whole EV drivetrain; while some other fault types might get escalated to a catastrophic EV drivetrain failure. In general, electric faults in higher power motors and inverters are more severe. A high number of drivetrain components increases the possibility of EV drivetrain failure. In concepts 4 and 5, if one of electric motors or inverters stops working, the vehicle might still be operational using other motors till it reaches for service.

The AC three-phase cables should be as short as possible (recommended less than 600 mm). Therefore, most inverters are located near electric motors. Generated EMI may directly affect other electronic controllers inside the vehicle. For in-wheel motors, the inverter needs to be near the tire that increases the generated EMI effect owing to the confined space. Noise control methods are complex and challenging in EVs owing to the complicated EMI noise propagation route [13].

All of the above comparison discussions are summarized in Table 1. As it can be seen, concept 4 (Two By-Wheel or Hub Motors + Single Ratio Gearbox) has the highest number that shows it is the most suitable for heavy-duty EV applications. Multiple manufacturers have developed electric drive axles (including suspension system) same as concept 4 for e-buses and trucks. ZF electric axle for buses and Dana electric axle for trucks are examples of this. The next two configurations that achieved same numbers are concept 1 (Central motor + Single Ratio Gearbox + Differential) and concept 3 (A Central Motor + A Differential). Various e-bus and trucks manufacturers have used both concepts; however concept 1 is more feasible owing to the market availability of their electric motors. Motors with high torque and low speed (suitable for concept 3) are not that much available in the market.

Table 1 Comparison of various EV drivetrain configurations.

Criteria	Concept 1		Concept 2		Concept 3		Concept 4		Concept 5	
Highest priority category		***X3***		***X3***		***X3***		***X3***		***X3***
High efficiency	3	9	2	6	3	9	3	9	5	15
Modularity	4	12	4	12	3	9	5	15	2	6
Low mass/volume	3	9	3	9	2	6	4	12	4	12
Low drivetrain cost	3	9	2	6	2	6	3	9	2	6
Low servicing time and cost	3	9	2	6	4	12	4	12	2	6
Control simplicity	4	12	2	6	5	15	3	9	2	6
Moderate priority category		***X2***		***X2***		***X2***		***X2***		***X2***
Retrofitting simplicity	3	6	3	6	4	8	3	6	2	4
Ease of servicing	3	6	3	6	4	8	3	6	2	4
Robustness and fault handling	5	10	2	4	2	4	5	6	3	6
Compliances/EMI	4	8	5	10	3	8	3	6	1	2
Least priority category		***X1***		***X1***		***X1***		***X1***		***X1***
Low acoustic noise	2	2	2	2	3	3	4	4	5	5
Low vibration	2	2	1	1	3	3	4	4	5	5
Overall addition	39	**92**	31	**74**	38	**91**	44	**98**	35	**77**

2.4 Chapter summary

Various heavy-duty EV drivetrain configurations have been explained, and their advantages and disadvantages have been discussed. These configurations have been compared in the context of EV design and development. Comparison studies show that concept 4 (Two By-Wheel or Hub Motors + Single Ratio Gearbox) and concept 3 (A Central Motor + A Differential) are the most suitable for heavy-duty EV applications accordingly.

References

[1] K. Bergsson, Hybrid Vehicle History More Than a Century of Evolution and Refinement, 2005. http://www.hybrid-vehicle.org/hybrid-vehicle-History.html. (Accessed 19 January 2020).
[2] K. Rajashekara, History of electric vehicles in general motors, IEEE Trans. Ind. Appl. 30 (1994) 897–904.
[3] A. Tashakori, M. Ektesabi, N. Hosseinzadeh, Characteristics of suitable drive train for electric vehicle, in: Proceeding of the International Conference on Instrumentation, Measurement, Circuits and Systems (ICIMCS 2011), Hong Kong, 2011, pp. 535–541.
[4] A. Tashakori, BLDC Motor Drives Controller for Electric Vehicles, (PhD thesis), May, 2014.
[5] F.A. Barata, J.C. Quadrado, J.F. Silva, Brushless DC motor: position linear control simulation, in: Proceedings of the 9th WSEAS International Conference on Systems, ICS'05 (Stevens Point, Wisconsin, USA), World Scientific and Engineering Academy and Society (WSEAS), 2005.
[6] X. Xue, K.W.E. Cheng, N.C. Cheung, Selection of electric motor drives for electric vehicles, in: Australian Universities Power Engineering Conference, AUPEC, 2008.
[7] O.C. Duffy, G. Wright, Medium/Heavy Truck Tasksheet Manual for NATEF Proficiency: 2014 NATEF; 2014.
[8] A. Tashakori, M. Ektesabi, Comparison of different PWM switching modes of BLDC motor as drive train of electric vehicles, World Acad. Sci. Eng. Technol. 67 (2012) 719–725.
[9] M. Ehsani, Y. Gao, S. Gay, Characterization of electric motor drives for traction applications, in: IECON Proceedings (Industrial Electronics Conference), 2003, Roanoke, VA, vol. 1, 2003, pp. 891–896.
[10] H. Tischmacher, I. Tsoumas, B. Eichinger, U. Werner, Case studies of acoustic noise emission from inverter-fed asynchronous machines, IEEE Trans. Ind. Appl. 47 (5) (2011) 2013–2022.
[11] P. Maloney, Electric Buses for Mass Transit Seen as Cost Effective, 2019. https://www.publicpower.org/periodical/article/electric-buses-mass-transit-seen-cost-effective. October 17 (Accessed 7 June 2020).
[12] California Air Resources Board, Literature Review on Transit Bus Maintenance Cost, California Air Resources Board, Sacramento, CA, 2016.
[13] N. Mutoh, M. Kanesaki, A suitable method for eco-vehicles to control surge voltage occurring at motor terminals connected to PWM inverters and to control induced EMI noise, IEEE Trans. Veh. Technol. 57 (4) (2008) 2089–2098.

CHAPTER 3

Electric Motor Drives for Heavy-duty Electric Vehicles

3.1 Introduction

Required power, operating voltage and current, the motor torque versus speed characteristics, control complexity, and cost are parameters that must be considered in selecting a suitable electric vehicle (EV) drivetrain for heavy-duty electric vehicles. These requirements have been discussed in more details in Chapter 2. Various electric motors have been used in heavy-duty applications so far. There are few research studies to compare the performance of various electric motors in heavy-duty electric vehicle applications. In this chapter, components in an electric motor drive and various electric motor types and their control strategies are discussed. Also a step-by-step guideline is given to calculate the required EV drivetrain power according to the vehicle specification.

3.1.1 Components in an electrical motor drive

It is assumed that the reader has a background understanding of electrical drives, but basic principles are briefly described. We represent the electrical drive as an electromechanical power converter, including an inverter drive and an electric motor. The motor may be turned to a generator by the inverter drive controller allowing kinetic energy recovery back to the battery. This functionality is commonly known as regenerative braking. The role of the inverter drive (or frequency converter) is to convert the direct current (DC) power into the alternating current (AC) power, generating the torque on the electrical motor rotor. This power conversion is reversed when the motor acts as a generator. The inverter adjusts and controls the rotation speed by controlling the AC voltage and frequency. The generated torque is controlled through modulation of the AC current going to the motor. The modern EV inverter drives are controlled by the vehicle control unit through CAN bus communication. As an example, the schematic

Heavy-duty Electric Vehicles
https://doi.org/10.1016/B978-0-12-818126-3.00006-3

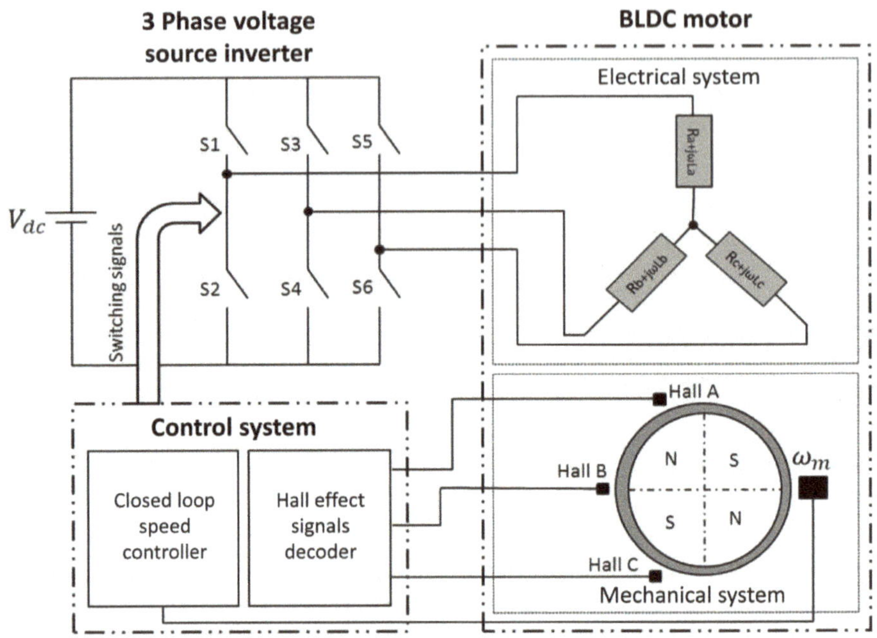

Fig. 1 Schematic diagram of a four-pole BLDC motor drive [1].

diagram of a four-pole permanent magnet brushless DC (BLDC) motor drive using Hall Effect Encoders is shown in Fig. 1 [1].

As can be seen from Fig. 1, there is a DC power supply input to the inverter, and the inverter drive includes six transistors (S1-S6) that are usually metal-oxide-silicon field-effect transistors (MOSFETs) or insulated gate bipolar transistors (IGBTs) for three-phase motors. The transistors' switching signals are controlled according to the feedback from the motor. The control strategies can differ depending on the type of motors.

3.1.2 Electrical motor types

The electrical motor consists of the stator and rotor. The stator has the windings conducting the electrical current, controlled by the inverter. The rotor of an electric motor will start to rotate once it experiences a changing magnetic field. The magnetic field is induced by the stator windings and rotates around the rotor's central axis in the vast majority of electrical motors. The rotor is supported by bearings usually connected to the stator frame to minimize vibrations.

The vast majority of the electrical motors are of radial-flux configuration, meaning the magnetic field flows radially from stator to rotor and back to the stator. The radial-flux motor configuration refers to an arrangement where

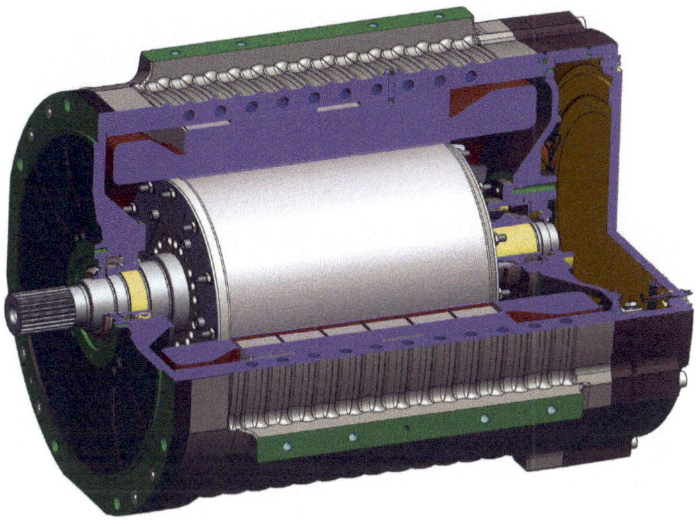

Fig. 2 A radial-flux motor CAD model. *(Courtesy of ABB.)*

the rotor and stator are co-centric, and the rotor has been enclosed within the stator. Radial-flux motor CAD model is shown in Fig. 2. As can be seen in the figure, the rotor, in gray, is supported by bearings (yellow). The power output on the left blue intersection shows the stator windings and stator core with cooling channels.

Outer rotor machines exist as well in some high-performance applications, for example, in drones. In the axial-flux electrical motors, the rotor and stator surfaces are co-centric as well but are located next to each other along the revolution axis. The magnetic flux flows axially from the stator to the rotor and back to the stator. Axial-flux motors have been introduced in vehicle applications as well [2].

The radial and axial flux configurations cover most of the rotating electrical machines used in industries (in addition to linear electrical machines). However, various innovative configurations have been presented, to mention, transversal flux motors [3], etc. It has been shown that motor type optimality is very case-specific. Regardless, the traditional radial-flux motors dominate the market [4].

The electric torque production process depends on the electrical motor type. The common principle is based on making the rotor to follow the rotating magnetic field. In the induction motors, the rotor is magnetized by the action of the rotating magnetic field induced by the stator coils. The field of the magnetized rotor produces a torque when following the rotating field.

Permanent magnet synchronous motors (PMSMs) have been used in the industrial application for the past 50 years. Nowadays, they are prevalent in automotive and aerospace fields. PMSMs are mainly of two types, based on their radial-flux motors [electromotive force (EMF)] waveform. Motors that have a sinusoidal back-EMF and motors that have a trapezoidal back-EMF are called BLDC Motors [1]. In PMSMs, the permanent magnets in the rotor form a magnetic field that follows the rotating magnetic field.

Switched reluctance motors (SRMs), also known as the variable reluctance motors, were built first in the 18th century. SRM's structure is similar to that of the PMSMs, but it has a ferromagnetic rotor; therefore it is cheaper. SRM's stator winding has electrically independent phases and requires an electronics commutation similar to PMSMs. Energizing the winding moves the rotor in a direction that minimizes the reluctance in the air gap and increases the stator winding inductance (when the rotor pole is aligned with stator pole), which results in generating the torque [1]. Permanent magnet and induction motors are the most common motor types in EVs.

A nonbiased comparison of different motor types is challenging because of multiple free parameters and matching a motor and drive with a purpose [5]. The most suitable type of motor depends on vehicle usage, life cycle, and manufacturing cost. A proper selection requires a profound techno-economical study taking into account vehicle specification, drive manufacturing cost, material cost, and life cycle cost. As a practical example, selecting a lower efficiency motor requires a slightly higher rating in the drive and in the battery. It would also lead to higher amount of energy consumed over the vehicle life cycle. Fig. 3 shows the schematic efficiency curves of three motor types. As it can be seen,

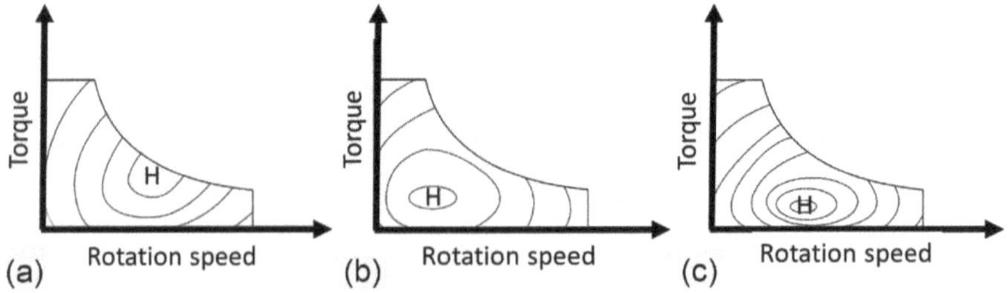

Fig. 3 Examples of efficiency charts of different electric motors as functions of speed and torque [5]. Letter H indicates high-efficiency area. (a) Permanent-magnet machine with SPM; (b) permanent-magnet machine with embedded magnets (IPM); and (c) induction machine.

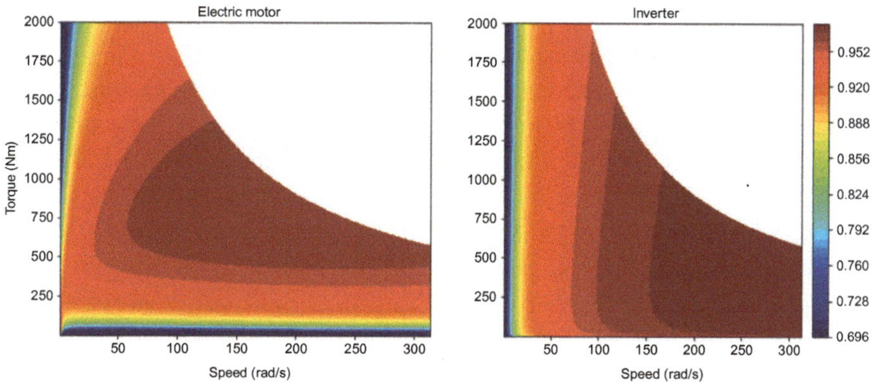

Fig. 4 Efficiency maps of electric motor and inverter as function of shaft speed and torque [7].

Surface-mounted magnets (SPMs) are more efficient in higher RMPs and higher torques, while IPMs are more efficient in lower torque and lower speeds operating points. Also, as shown, the induction motor's high-efficient area is much smaller than permanent-magnet motors.

A comparison study shows that permanent magnet synchronous motors are suitable for EVs because of their high efficiency, fast dynamic torque response, and high power/energy to size ratio [6]. Consequently, it is the most common motor type in commercial EVs. Fig. 4 shows efficiency maps of electric motor and inverter as function of shaft speed and torque [7, 8]. It should be noted that efficiency maps are approximations based on semianalytical equations. These approximations are likely sufficient for feasibility studies but should not be considered exact for any electrical machine. Dynamometric measurements are required to calculate efficiency, and calorimetric measurements are required to have exact loss profile for a given machine.

3.1.3 Motor inverter drives

The high-level task of an inverter is to convert the DC voltage and current to AC, which in turn creates a time-changing magnetic field in the electrical machine. The power stage in the inverter typically is an H-bridge amplifier (as shown in Fig. 1), driven by pulse width modulation (PWM) to regulate the power and frequency. Transistors S1-S6 connect phases either to the maximum voltage or to the minimum voltage. Note that, for example, switches S1 and S2 must not be coupled simultaneously because it leads to a short circuit in the battery. The voltage at the inverter output is of rectangular type, but the current is close to sinusoidal because it is filtered by the windings in electrical machine.

Inverters manufactured for EV applications must be able to control the motor torque at slow speeds. For instance, no bumping at the front of a pedestrian way is allowed. The other challenging operation mode is field weakening. The converter must keep its sync with a motor. Losing the sync might mean locking the motor and/or blowing the power stage in the inverter. These days, silicon-based IGBT power stages are prevalent; however, research work is on the Gallium-Nitride-based devices and the Silicon Carbide MOSFETs. Silicon Carbide MOSFETs are found to improve the efficiency and power density of power stages [9, 10].

3.2 Power consumption in heavy-duty electric vehicles

The driving cycle and auxiliary loads need to be defined or assumed in order to define the total power and energy consumption. In this section, we focus on the EV drivetrain load to propel the wheels. However, in some applications, continuous power consumption by the vehicle auxiliary loads might be larger than propulsion loads in certain operating conditions. For instance, air conditioning or heating of the vehicle cabin may take more power than the average propulsion load [11]. Naturally, the battery or other power sources must serve the total load. Another note is that kinematics-based approach can serve for worst-case studies and predetermined cases. The optimal design could be achieved by analyzing the actual driving cycles and by considering life cycle efficiency [12].

3.2.1 EV drivetrain power calculations

The first step to select the suitable EV drivetrain motor and inverter is to know the needed drivetrain power; battery voltage, and energy based on the vehicle performance requirements. Fig. 5 shows the heavy EV body on a slope and applied forces to it.

The rigid body movement of the vehicle on a road is described with the following Eqs. (1), (2), where the inertia of the rotating components (wheels and shafts) of the vehicle is neglected to simplify the calculations.

$$m\frac{dv_x}{dt} = \Sigma\left(F_x - F_d - mg\sin(\beta)\right) \tag{1}$$

$$F_d = \frac{1}{2}C_d\rho A v_x^2 + f_{RR}mg\cos(\beta) \tag{2}$$

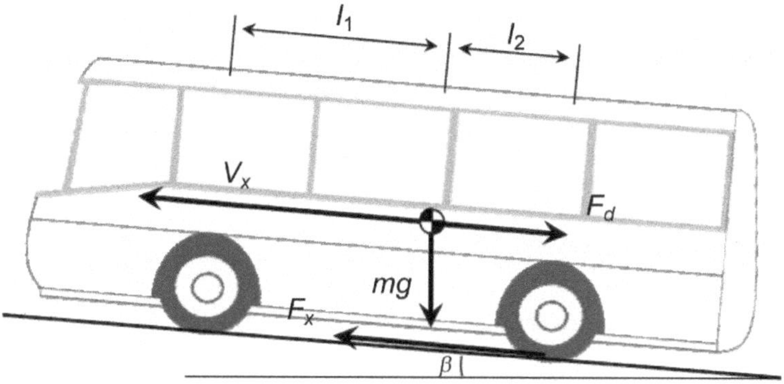

Fig. 5 The heavy electric vehicle body on a slope and applied forces.

Converting the tractive force to the motor torque is done with the help of Eq. (3), where the slip between the wheels and road is neglected to simplify the calculations.

$$F_x \frac{D_{dyn}}{2} = T_w = T_e i \tag{3}$$

Note that the transmission is assumed to be a fixed speed transmission with no losses. A simple overall design procedure is needed to determine first the power rating of various components in EV powertrain. After the power rating is fixed, the battery voltage level can be iterated. It is worth noting that the practical design task often takes into account earlier designs, company tradition, suppliers' offering, etc. This design procedure omits those aspects that elaborate physical boundaries of the powertrain design.

In the EV design, battery voltage, electric drive maximum current, and gear ratio are coupled to the vehicle traction force and speed. Therefore a change, for example, in gear ratio may influence on battery voltage selection. The power balance of powertrain design can be achieved by the following iteration:

1. Determine driving cycles (use the rigid body equations but preferably multiple driving cycle profiles to study the setup).
2. Determine loads (in multiple environments).
3. Compute motor-generator power (estimate the speed-power curve for the vehicle).
4. Compute the inverter power.
5. Compute battery power and capacity required (remember losses between road and battery).
6. Compute battery power and capacity from charging requirement (remember losses between grid and battery).

Once the power balance is achieved, possibly after iterations, the potential and flow variables (meaning voltage and current, speed, and torque) can be fixed. If a reader is interested in generalized potential and flow variables, we recommend familiarizing with, for example, energetic macroscopic representation (EMR) [12]. We do not use EMR in this book but acknowledge it as a formalized method for modeling and understanding mechatronic power systems. We iterate the powertrain design by the following steps:

1. Fix battery voltage first (consider charging too).
2. Compute battery current (remember losses and check the feasibility of cable thicknesses).
3. Fix gear ratio (EVs often have one fixed gear ratio).
4. Compute motor speed and torque (at different grades).
5. Compute motor and inverter current.
6. If the design is not feasible, return to 1. Increasing battery voltage and maximum motor speeds will reduce the current and resistive losses in cabling.

In the time of writing this book, practical battery voltages for IGBT power stages were in a range of 400–800 V, and maximum motor speeds were in a range of 6000–15,000 rpm. Light vehicle drives may soon approach 20,000 rpm, but heavy vehicle drives operate at a lower speed.

In case the design is not achievable at all, the powertrain components may be multiplied by designing parallel powertrains within the same vehicle. Multiplication of drives is natural for four-wheel-, or six-wheel-driven vehicles, for instance. Using an electric drive per axle is a common approach already today. Some high performance vehicles may use an electric drive for each wheel. Please refer to Chapter 2 for more information on EV drivetrain configurations.

3.3 Case study

3.3.1 Power and torque calculations

As a case study example, the required power, torque, and speed values are calculated for an electric bus with a central powertrain and given parameters in Table 1.

The required motor torque and speed calculations have been done for speeds up to 100 km/h and for road grades up to 12%. These values are selected based on the information available in Chapter 2. The calculation results are given in Table 2. The results show vehicle needs 188.2 kW of electrical power to run 100 km/h on a flat road. Also, as can be seen, the

Table 1 Case study—Electric bus specifications.

Parameter	**Value**
Kerb weight (kg)	12,500
Payload (kg)	4000
GVM (kg)	16,500
f_{RR} (coefficient of RR)	0.02
C_d (drag coefficient)	0.6
A (frontal area—m^2)	9
ρ (density of air—kg/m^3)	1.2
g (gravity—m/s)	9.81
V (final velocity—km/h)	100
θ (gradient—%)	0–12
R (wheel radius—m)	0.286
Transmission ratio	2.92
Transmission efficiency	0.97
Tire efficiency	0.99
Total powertrain efficiency	0.9

required power increases exponentially as the vehicle speed and road grade increase. The 3D plot of the required power at different speeds and grades are shown in Fig. 6.

The next step is to select a motor that can deliver the calculated torque at the given speed and grade. Electric buses usually must be able to run 100 km/h on a flat road, and required speed on various grades depends on the operating drive cycle. By comparing the selected motor torque-speed graph and the calculated values, we can find operation limits of the vehicle. For example, Fig. 7 shows the required torque calculations at different grades and the SUMO MD MV2500-6P performance graph manufactured by Dana Tm4. As can be seen, the required torque and speed of this electric bus on a flat road (at 0% grade) is within operational limits of the motor. However, as the road grade increases, then the available speed operation limit of the vehicle decreases. As it can be seen, the SUMO MD MV2500-6P motor is a perfectly suitable choice for our case study application.

Table 2 Case study—Torque and speed calculations.

Grade (%)	Vehicle speed (km/h)	f_{RR} (rolling resistance force—N)	F_{ad} (aero drag force—N)	Drag power (kW)	Gravity force (N)	Gravity power (kW)	Total power (kW)	Torque wheel (Nm)	Motor torque (Nm)	Motor speed (RPM)	Motor power (kW)	Electrical power (kW)
0	5	3237.3	6.25	4.50	0.00	0.00	4.50	956.35	337.64	135.41	4.79	5.32
3	5	3237.3	6.25	4.50	4853.77	6.74	11.25	2387.46	842.91	135.41	11.95	13.28
6	5	3237.3	6.25	4.50	9694.47	13.46	17.97	3814.71	1346.81	135.41	19.10	21.22
9	5	3237.3	6.25	4.50	14509.21	20.15	24.66	5234.32	1848.02	135.41	26.21	29.12
12	5	3237.3	6.25	4.50	19285.44	26.79	31.29	6642.57	2345.21	135.41	33.26	36.95
0	10	3237.3	25.00	9.06	0.00	0.00	9.06	961.87	339.60	270.82	9.63	10.70
3	10	3237.3	25.00	9.06	4853.77	13.48	22.54	2392.98	844.86	270.82	23.96	26.62
6	10	3237.3	25.00	9.06	9694.47	26.93	35.99	3820.24	1348.77	270.82	38.25	42.50
9	10	3237.3	25.00	9.06	14509.21	40.30	49.37	5239.85	1849.97	270.82	52.47	58.30
12	10	3237.3	25.00	9.06	19285.44	53.57	62.63	6648.10	2347.16	270.82	66.57	73.96
0	20	3237.3	100.00	18.54	0.00	0.00	18.54	983.99	347.40	541.65	19.71	21.89
3	20	3237.3	100.00	18.54	4853.77	26.97	45.51	2415.10	852.67	541.65	48.36	53.74
6	20	3237.3	100.00	18.54	9694.47	53.86	72.40	3842.36	1356.57	541.65	76.95	85.50
9	20	3237.3	100.00	18.54	14509.21	80.61	99.15	5261.96	1857.77	541.65	105.37	117.08
12	20	3237.3	100.00	18.54	19285.44	107.14	125.68	6670.21	2354.97	541.65	133.58	148.42

0	30	3237.3	225.00	28.85	0.00	0.00	28.85	1020.84	360.42	812.47	30.66	34.07
3	30	3237.3	225.00	28.85	4853.77	40.45	69.30	2451.95	865.68	812.47	73.65	81.84
6	30	3237.3	225.00	28.85	9694.47	80.79	109.64	3879.21	1369.58	812.47	116.53	129.47
9	30	3237.3	225.00	28.85	14509.21	120.91	149.76	5298.82	1870.79	812.47	159.17	176.85
12	30	3237.3	225.00	28.85	19285.44	160.71	189.56	6707.07	2367.98	812.47	201.47	223.86
0	40	3237.3	400.00	40.41	0.00	0.00	40.41	1072.44	378.63	1083.29	42.95	47.73
3	40	3237.3	400.00	40.41	4853.77	53.93	94.35	2503.55	883.90	1083.29	100.27	111.41
6	40	3237.3	400.00	40.41	9694.47	107.72	148.13	3930.81	1387.80	1083.29	157.44	174.93
9	40	3237.3	400.00	40.41	14509.21	161.21	201.63	5350.41	1889.00	1083.29	214.29	238.10
12	40	3237.3	400.00	40.41	19285.44	214.28	254.70	6758.66	2386.20	1083.29	270.70	300.77
0	50	3237.3	625.00	53.64	0.00	0.00	53.64	1138.78	402.06	1354.12	57.01	63.35
3	50	3237.3	625.00	53.64	4853.77	67.41	121.06	2569.89	907.32	1354.12	128.66	142.96
6	50	3237.3	625.00	53.64	9694.47	134.65	188.29	3997.15	1411.22	1354.12	200.12	222.35
9	50	3237.3	625.00	53.64	14509.21	201.52	255.16	5416.75	1912.43	1354.12	271.19	301.32
12	50	3237.3	625.00	53.64	19285.44	267.85	321.50	6825.00	2409.62	1354.12	341.69	379.66
0	60	3237.3	900.00	68.96	0.00	0.00	68.96	1219.86	430.68	1624.94	73.29	81.43
3	60	3237.3	900.00	68.96	4853.77	80.90	149.85	2650.97	935.95	1624.94	159.26	176.96
6	60	3237.3	900.00	68.96	9694.47	161.57	230.53	4078.23	1439.85	1624.94	245.01	272.23
9	60	3237.3	900.00	68.96	14509.21	241.82	310.78	5497.84	1941.05	1624.94	330.30	367.00

Continued

Table 2 Case study—Torque and speed calculations—cont'd

Grade (%)	Vehicle speed (km/h)	f_{RR} (rolling resistance force—N)	F_{ad} (aero drag force—N)	Drag power (kW)	Gravity force (N)	Gravity power (kW)	Total power (kW)	Torque wheel (Nm)	Motor torque (Nm)	Motor speed (RPM)	Motor power (kW)	Electrical power (kW)
12	60	3237.3	900.00	68.96	19285.44	321.42	390.38	6906.09	2438.25	1624.94	414.90	461.00
0	70	3237.3	1225.00	86.77	0.00	0.00	86.77	1315.69	464.51	1895.76	92.22	102.46
3	70	3237.3	1225.00	86.77	4853.77	94.38	181.15	2746.80	969.78	1895.76	192.52	213.92
6	70	3237.3	1225.00	86.77	9694.47	188.50	275.27	4174.06	1473.68	1895.76	292.56	325.07
9	70	3237.3	1225.00	86.77	14509.21	282.12	368.89	5593.66	1974.88	1895.76	392.06	435.62
12	70	3237.3	1225.00	86.77	19285.44	374.99	461.76	7001.91	2472.08	1895.76	490.77	545.30
0	80	3237.3	1600.00	107.50	0.00	0.00	107.50	1426.26	503.55	2166.58	114.25	126.94
3	80	3237.3	1600.00	107.50	4853.77	107.86	215.36	2857.37	1008.81	2166.58	228.88	254.32
6	80	3237.3	1600.00	107.50	9694.47	215.43	322.93	4284.62	1512.72	2166.58	343.21	381.35
9	80	3237.3	1600.00	107.50	14509.21	322.43	429.92	5704.23	2013.92	2166.58	456.93	507.70
12	80	3237.3	1600.00	107.50	19285.44	428.57	536.06	7112.48	2511.11	2166.58	569.73	633.04
0	90	3237.3	2025.00	131.56	0.00	0.00	131.56	1551.56	547.79	2437.41	139.82	155.36
3	90	3237.3	2025.00	131.56	4853.77	121.34	252.90	2982.68	1053.06	2437.41	268.79	298.65

6	90	3237.3	2025.00	131.56	9694.47	242.36	373.92	4409.93	1556.96	2437.41	397.41	441.56
9	90	3237.3	2025.00	131.56	14509.21	362.73	494.29	5829.54	2058.16	2437.41	525.33	583.71
12	90	3237.3	2025.00	131.56	19285.44	482.14	613.69	7237.79	2555.36	2437.41	652.24	724.71
0	100	3237.3	2500.00	159.37	0.00	0.00	159.37	1691.62	597.24	2708.23	169.38	188.20
3	100	3237.3	2500.00	159.37	4853.77	134.83	294.20	3122.73	1102.50	2708.23	312.68	347.42
6	100	3237.3	2500.00	159.37	9694.47	269.29	428.66	4549.98	1606.41	2708.23	455.59	506.21
9	100	3237.3	2500.00	159.37	14509.21	403.03	562.40	5969.59	2107.61	2708.23	597.73	664.14
12	100	3237.3	2500.00	159.37	19285.44	535.71	695.08	7377.84	2604.80	2708.23	738.74	820.82

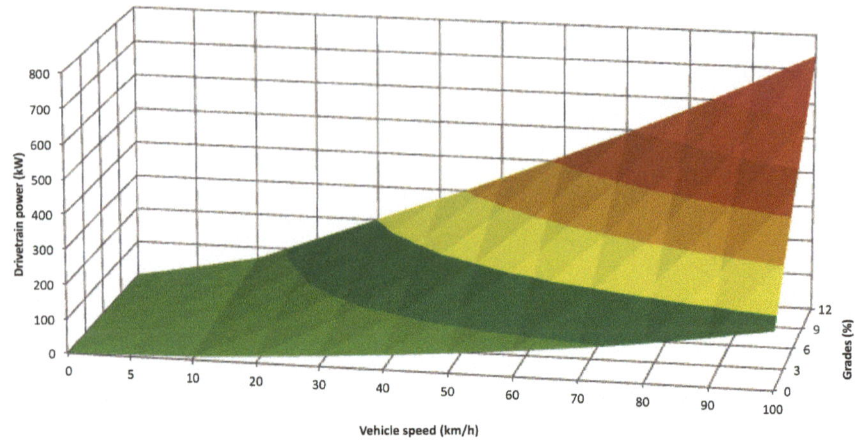

Fig. 6 The 3D plot of the calculated power at different speeds and grades.

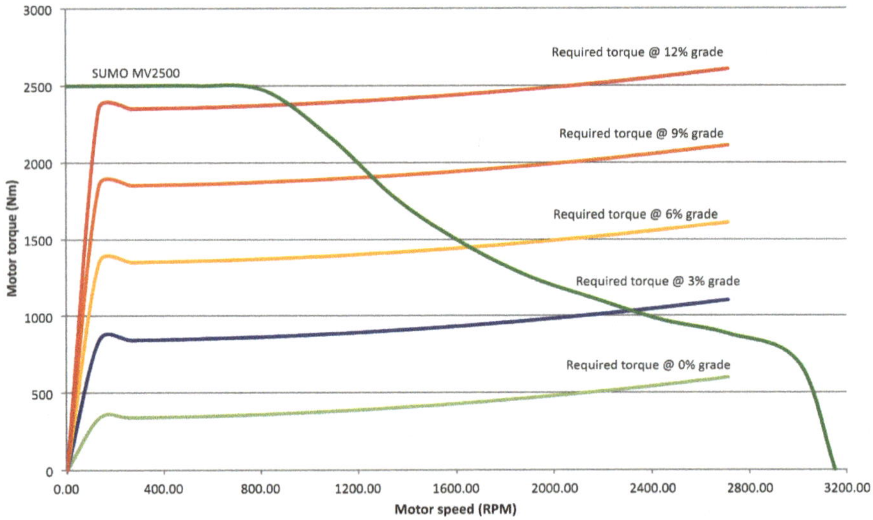

Fig. 7 The required torque calculations at different grades and the SUMO MD MV2500-6P performance graph manufactured by Dana Tm4.

3.3.2 Electrical vehicle system level model

A simplified simulation model of a heavy-duty vehicle with the mass 10,000 kg is shown in Fig. 8. The vehicle performance is simulated using NEDC (New European Drive Cycle). The A proportional-integral-derivative (PID) speed controller performs the vehicle speed control, and the reference speed profile input is representative of the driver demand.

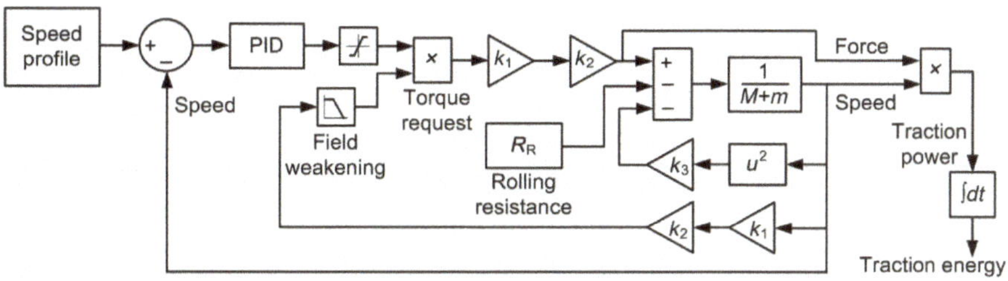

Fig. 8 A heavy-duty vehicle simulation model created in Matlab/Simulink [13].

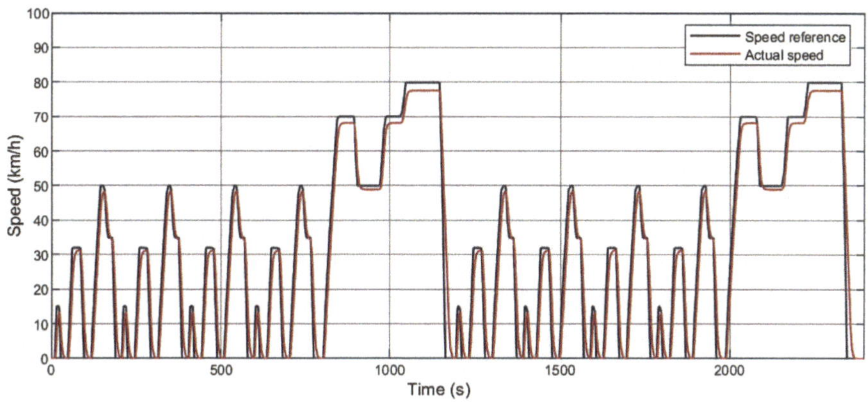

Fig. 9 Actual vehicle speed and the speed command.

The propulsion motor has only a few nonidealities: limited torque production capability as a function of speed and limited speed. The reduction gear has the gear ratio of 8 and ideal efficiency. The drag forces are being subtracted from propulsion force before sending the resultant force in the inertia named "Vehicle body, mass." The traction force, power, and energy are measured. It is notable that the model has numerous limitations and idealizations. However, this model is a starting point for many analyses, and we believe it is pedagogically sufficiently simple.

The actual vehicle speed and the speed command are shown in Fig. 9. It can be seen that the actual vehicle speed follows the speed command. The drivetrain power of the vehicle is shown in Fig. 10. As can be seen, drivetrain power consumption reaching nearly 100 kW with multiple regenerative power notches owing to deceleration/braking.

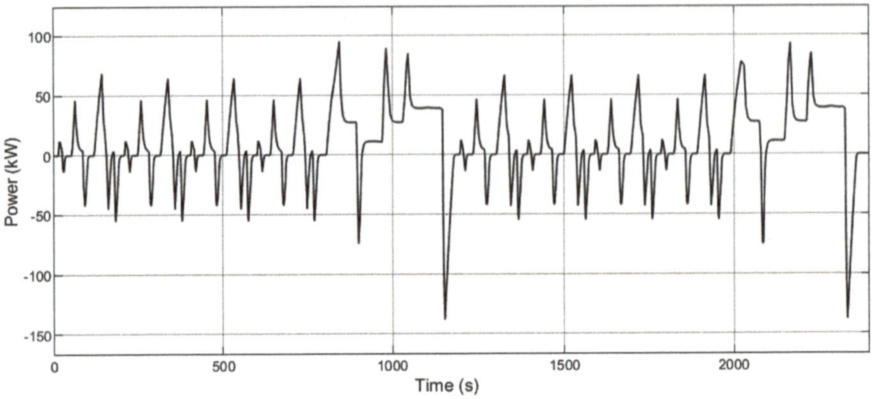

Fig. 10 Vehicle drivetrain power.

3.4 Summary

Basic concepts of EV drivetrain components, including electric motors and inverters, are discussed in this chapter. Various motor types and their characteristics, such as operation principle and efficiency maps, are reviewed. Step-by-step required power calculation for a heavy-duty EV is explained and validated through a case study. Results are compared against the selected electric motor torque/speed characteristics to analyze the EV drivetrain performance capability. Also a heavy-duty EV model is created in Simulink, and the vehicle performance is presented for the New European Drive Cycle.

References

[1] A. Tashakori, BLDC Motor Drives Controller for Electric Vehicles (PhD thesis), Swinburne University of Technology, Melbourne, Australia, May 2014.

[2] A. Gu, B. Ruan, W. Cao, Q. Yuan, Y. Lian, H. Zhang, A general SVM-based multi-objective optimization methodology for axial flux motor design: YASA motor of an electric vehicle as a case study, IEEE Access. 7 (2019) 180251–180257.

[3] G. Henneberger, M. Bork, Development of a new transverse flux motor, in: New Topologies for Permanent Magnet Machines (Digest No: 1997/090), IEE Colloquium, 1997, pp. 1/1–1/6.

[4] J. Pippuri, A. Manninen, J. Keränen, K. Tammi, Torque density of radial, axial and transverse flux permanent magnet machine topologies, IEEE Trans. Magn. 49 (5) (2013). 4 p.

[5] J. De Santiago, H. Bernhoff, B. Ekergård, S. Eriksson, S. Ferhatovic, R. Waters, M. Leijon, Electrical motor drivelines in commercial all-electric vehicles: a review, IEEE Trans. Veh. Technol. 1939-9359, 61 (2) (2012) 475–484, https://doi.org/10.1109/TVT.2011.2177873.

[6] A. Tashakori, M. Ektesabi, N. Hosseinzadeh, Characteristics of suitable drive train for electric vehicle, in: Proceeding of the International Conference on Instrumentation, Measurement, Circuits and Systems (ICIMCS 2011), Hong Kong, December 2011, pp. 535–541.
[7] A. Ritari, J. Vepsäläinen, K. Kivekäs, K. Tammi, H. Laitinen, Energy consumption and lifecycle cost analysis of electric city buses with multispeed gearboxes, Energies (2020). 21 p.
[8] A. Mahmoudi, W.L. Soong, G. Pellegrino, E. Armando, Efficiency maps of electrical machines, in: 2015 IEEE Energy Conversion Congress and Exposition (ECCE), Montreal, QC, 2015, pp. 2791–2799.
[9] T. Bertelshofer, R. Horff, A. Maerz, M.-M. Bakran, A performance comparison of a 650 V Si IGBT and SiC MOSFET inverter under automotive conditions, PCIM Europe. International Exhibition and Conference for Power Electronics, Intelligent Motion, Renewable Energy and Energy Management, 2016.
[10] X. Ding, M. Du, T. Zhou, H. Guo, C. Zhang, F. Chen, Comprehensive comparison between sic-mosfets and si-igbts based electric vehicle traction systems under low speed and light load, in: CUE2015-Applied Energy Symposium and Summit 2015: Low Carbon Cities and Urban Energy Systems, 2015.
[11] K. Kivekäs, J. Vepsäläinen, F. Baldi, A. Lajunen, K. Tammi, Reducing the energy consumption of electric buses with design choices and predictive driving, IEEE Trans. Veh. Technol. 68 (12) (2019) 11409–11419.
[12] K.S. Agbli, D. Hissel, M.-C. Péra, I. Doumbia, Energetic macroscopic representation (EMR): new approach for multiphysics energetic flows modelling, IFAC Proc. Vol. 45 (21) (2012) 723–728.
[13] J. Vepsäläinen, K. Otto, A. Lajunen, K. Tammi, Computationally efficient model for energy demand prediction of electric city bus in varying operating conditions, Energy 169C (2019) 433–443.

PART II

Energy Storage and Charging

CHAPTER 4

Materials and Manufacturing Methods for Advanced Li-ion Batteries

4.1 Introduction

In their present form and configuration, lithium-ion (Li-ion) batteries offer energy storage capacities ranging between 110 and 180 $Whkg^{-1}$. This is up to four times lower than the capacity available at material's level owing to the presence of inactive components such as current collectors, additives, electric wiring and power electronics, and packaging. For the same reason, volumetric energy density at the material's level is also approximately two times higher than at a cellular level [1–3]. Increasing attractiveness of long-range electric vehicles (EVs) has created a demand for new generation Li-ion batteries [4].

The new generation batteries are required to have high energy and power densities to maximize the usable space and the load-carrying capacity of EVs [5–7]. Because electrical energy stored in a Li-ion battery is a function of its operating voltage and nominal capacity, the impetus is on developing Li-ion battery systems with high terminal voltage. One method of achieving this is through replacement of traditional cathodes, such as spinel lithium manganese oxide and layered lithium cobalt oxide, with active materials that exhibit higher redox potentials than lithium reference electrode (Li^+/Li) [8]. As far as anodes are concerned, commercially implemented capacity for graphite, the most common carbon-based anodic material, is fast approaching its theoretical maximum limit. Silicon (Si), a naturally abundant mineral with a low operating potential (<0.5 V vs Li/Li^+) and the highest known theoretical capacity (4200 $mAhg^{-1}$), is considered a promising replacement for carbon-based anodes in high-energy Li-ion batteries. However, Si-anodes experience 100%–320% volumetric change during lithiation and delithiation cycles [9]. Some gains can also be made by reducing the percentage of inactive

Heavy-duty Electric Vehicles
https://doi.org/10.1016/B978-0-12-818126-3.00007-5

components. The approach may involve fabrication of binderless electrodes or thick electrodes. Inherent features of alternate battery manufacturing methods can support these goals.

This chapter presents a brief overview of various research efforts made to improve the characteristics and performance of Li-ion batteries in different operating conditions. First, developments made in the materials domain are described. The subsequent section presents some of the advanced manufacturing methods that can boost the development of next-generation batteries.

4.2 Battery materials

All batteries have three primary components: electrodes—positive electrode or cathode and negative electrode or anode, electrolyte (salt + solvent), and a separator. The positive electrode layer is coated over aluminum current collector foil, whereas a copper current collector is used as the substrate for negative electrodes. In addition, binder, conductive agent, and other additives such as flame retardants and gel precursors are added to improve the battery performance. Table 1 presents a list of typical materials used in making Li-ion batteries.

Table 1 Materials used for making Li-ion batteries.

Component		**Materials**
Anode		Natural graphite, synthetic graphite, hard carbon, LTO, silicon-based alloys, tin-based alloys
Cathode		LFP, LCO, LMO, NCA, NMC, ECPs
Electrolyte	Salt	$LiPF_6$, $LiBF_4$, $LiClO_4$, $LiCF_3SO_3$, $LiAsF_6$
	Solvent	EC, DMC, DEC, PC, RTIL, γ-GBL
Separator		Cellulosic paper, polyethylene, polypropylene, ceramics, nonwoven fabrics
Binder		PVDF, CMC, SBR, ACM, Glass fiber
Flame retardant		TFP, TMP, HMPN
Gel precursor		PVDF, PTFE, PAN, PEO, PMMA

ACM, acrylate copolymer; *ECPs*, electronic conducting polymers; γ-*GBL*, gamma-butyrolactone; *HMPN*, hexamethoxycyclotriphosphazene; *LTO*, lithium titanate oxide; *PAN*, polyacrylonitrile; *PC*, propylene carbonate; *PEO*, polyethylene oxide; *PMMA*, polymethyl methacrylate; *RTILs*, room temperature ionic liquids; *TFP*, tris(2,2,2-trifluoroethyl)phosphate; *TMP*, trimethyl phosphate [10] (Reprinted with permission).

4.2.1 Anode

Anodic reactions can be divided into three categories: intercalation reactions, conversion reactions, and alloying reactions. Correspondingly, there are three types of anode materials to support these reactions.

4.2.1.1 Intercalation anode materials

Graphite and LTO are the most common intercalation anode materials. State-of-the-art Li-ion batteries use carbonaceous anodes owing to their ready availability, favorable electrochemical properties, and low cost. However, graphite has a comparatively low theoretical lithium intercalation capacity of $372\,mAhg^{-1}$ and a poor Li-ion diffusion coefficient of the order of $10^{-9}\,cm^2s^{-1}$, limiting its applicability in the next-generation batteries. Another member of this category—amorphous carbons, created through pyrolysis of organic precursors, have a malleable microstructure and can support a much higher lithium uptake than graphitic anodes. Besides, varieties of artificial graphite have been synthesized, but their synthesis requires managing gases needed for heat treatment at temperatures exceeding 3000°C. Thus, it is preferred to modify the structure of natural graphite by blending it with forms such as kish graphite. Such surface and structural modifications have resulted in a 10% average increase in the energy density of Li-ion batteries for successive years [11]. Nevertheless, further improvements are needed to compete with increasing choices of alternate materials.

Graphene with its high electrical conductivity, superior mechanical strength, and a surface to mass ratio greater than $2600\,m^2g^{-1}$ has been proposed as a probable substitute for graphite anodes. In practice, a single layer of graphene is deposited onto a current collector foil by methods such as chemical vapor deposition. Owing to the small thickness of this layer, repulsive forces between the deposited Li-ions become dominant and limit Li-ion uptake capability. For this reason, chemically modified graphene, for example, graphene oxide (GO) is preferred over a single-layer graphene. Specific capacity of $100\,mAg^{-1}$ at the current rate of $29\,mAg^{-1}$ has been achieved for GO already at full-cell scale [12].

Graphene is also available in nanoflakes form. Graphene nanoflakes possess high crystallinity and can be mass-produced from pristine graphite through liquid exfoliation method. With respect to graphene basal plane, edges of the graphene nanoflakes are regarded as highly active Li-ion intercalation sites. They offer up to 50% higher binding energy (1.70–2.27 vs 1.55 eV) and an energy barrier smaller by up to 0.15 eV for Li-ion diffusion. Furthermore, the lateral size of nanoflakes used in anode manufacturing is a crucial parameter affecting battery performances. Graphene nanoflakes with a lateral size smaller than 100 nm provide a larger edge to the bulk ratio of carbon atoms. Theoretically, nanostructured anodes can have a significantly higher gravimetric density as opposed to a conventional graphitic anode [11].

Among the recently developed anode materials, $Li_4Ti_5O_{12}$ has attracted the most attention. It operates at a potential of 1.55 V, which inhibits dendritic formation during low temperature and during high discharge operation. $Li_4Ti_5O_{12}$ exists in a spinel-shaped crystal structure with Fd3m space grouping. This means that the three-dimensional (3D) structure remains unchanged during the electrochemical reactions, and the resultant volume changes are less than 0.1%. Crystal structure allows further modifications to enhance the charge transfer kinetics by doping it with conductive materials. Decreasing particle size is also a viable option. The chemistry is inherently safe with good cyclic stability. Moreover, the stability of $Li_4Ti_5O_{12}$ anodes does not depend on the presence of a passive surface film; their operating potential of 1.55 V is significantly higher than the redox voltages of the majority of commonly used electrolytes. As a result, they illustrate remarkable low temperature performance. $Li_4Ti_5O_{12}$ can also be employed in the form of a nanopowder, which is partly responsible for its excellent charge transfer kinetics and enables fast charging of electrodes. Nevertheless, small theoretical capacity (175 $mAhg^{-1}$) restricts their large-scale adoption by EV industries. Thus quest for better anode materials, that is, materials that offer comparable benefits to $Li_4Ti_5O_{12}$ along with a larger specific capacity, continues.

Han et al. reported an intercalation-type compound ($TiNb_2O_7$) featuring a layered monoclinic structure. Its crystal structure falls under C2/m space category. Edges and corners of the $TiNb_2O_7$ monoclinic crystal are shared by three ReO_3 blocks containing octahedral sites that are occupied by Nb_5 and Ti_4 ions in random order. Consequently, $TiNb_2O_7$ exhibits a more open crystal framework than LTO materials. It also facilitates safer operation than LTO anodes owing to a comparatively higher operating potential (1.60 vs 1.55 V). More importantly, energy capacities over two times greater than that of the conventional LTO anodes can be realized by replacing them with anodes made from $TiNb_2O_7$ material (388 vs 175 $mAhg^{-1}$), essentially owing to availability of five free electrons per formula unit. Limited charge transfer capability of $TiNb_2O_7$ is a problem although. Vacant 4d/3d orbitals render insulator-like characteristics to the $TiNb_2O_7$ crystal, which are noticeable in the form of its low electronic conductivity and ionic conductivity [13]. Doping could be a possible solution. Wen et al. used V_5 as a doping agent to improve Li-ion diffusion in the host material. They demonstrated a comparable capacity of 172 $mAhg^{-1}$ in $TiV_{0.02}Nb_{1.98}O_7$ at a 10C-rate [14]. Lin et al. managed to achieve a capacity of 181 $mAhg^{-1}$ at 5C using $Ru_{0.01}Ti_{0.99}Nb_2O_7$ [15]. On the other hand, Song and Kim preferred Mo_6 as a doping material and demonstrated an impressive energy storage capacity of (190 $mAhg^{-1}$) $Mo_{0.05}Ti_{0.95}Nb_2O_7$ at 10C-rate [16]. Although the research is promising, issues related to toxicity, high cost, and large molecular weight of the three aforementioned doping agents, respectively, limits the applicability of the doped variants. New doping materials and methods are, therefore, needed.

Carbon nanotubes (CNTs) have also received significant attention owing to their large specific surface area, high conductivity, and a unique one-dimensional microstructure. Multiwalled CNTs, in particular, provide a large number of Li-intercalation sites owing to local turbostratic disorder and a central core. In fact, Li-uptake capacities over 1400 $mAhg^{-1}$ are regularly reported in published literature. Moreover, CNTs are chemically stable and have high electrical conductivity and thermal conductivity, 10^{-4} Scm^{-1} and 2000–4000 $Wm^{-1}K^{-1}$, respectively. They also possess high shear strength ($\sim$ 500 MPa) in addition to excellent tensile strength ($\sim$ 50 GPa). Short Li-ion diffusion pathways created by virtue of the CNT microstructure enhance charge transfer dynamics for the system and enable charging/discharging at high rates. Durability, high strength, and significantly improved power density and energy density are anticipated to be key features of CNT-based anodes. However, similar to the most other novel anode materials, poor cyclic stability and a large irreversible capacity loss over time affect CNT-based anodes as well. Combining CNTs with Li-alloying metals has been proposed as a possible solution to the problem. For example, Kumar et al. demonstrated Sn-filled multiwalled CNT-based high-capacity Li-insertion anodes. De-insertion capacity of 889 $mAhg^{-1}$ was realized for the first cycle [17]. Chen et al. used aligned-CNT arrays to design anode for a Li-ion battery. Discharge capacity of the battery (over 265 $mAhg^{-1}$) remained stable for more than 50 cycles. An additional benefit of aligned CNT-based anode is that it made copper current collector redundant. This resulted in a weight saving of more than 85% [18]. Risk of potential battery failure owing to copper pitting and dissolution in the electrolyte was also eliminated using copper-less anode.

4.2.1.2 Alloying anode materials

Alloy anodes are fabricated either from pure metals, alloys, or intermetallic compounds. Both amorphous and crystalline states are acceptable in the form of fine powders and as thin films. The primary benefit of alloy anodes is that they provide 2–10 times higher theoretical specific capacities than the graphite anodes and 4–20 times higher specific capacities than the LTO anodes [19]. A moderate onset potential of 0.3–0.6 V above Li/Li^+ mitigates the risk of lithium deposition and offers another crucial advantage over graphite anodes. Energy penalty is also not as large as in the case of LTO anodes (1.5 V above Li/Li^+). They have thus been researched heavily in the past few years. The key elements are Si and Tin (Sn). Lithiation up to a stoichiometric ratio of 4.4 can be readily performed with both Si and Sn, which corresponds to specific capacities of 4200 and 994 $mAhg^{-1}$, respectively. Despite demonstrating better capacity retainability than plain Sn/Si electrodes, intermetallic anodes, for example, Ni_3Sn_4 are not considered suitable for prolonged cycling. Cycle life of these anodes is compromised by large volume changes (can be up to 320% in Si anodes) accompanying the alloying process. Mechanical stress created owing to volume expansion of anode particles results in the failure of active mass integrity, which causes capacity fade in batteries. Besides, the formation of Li-alloys in an electrochemical cell with Si-anode is followed by pulverization of electrode and, subsequently, loss of electrical contact between the Si-layer and the adjoining current collector [9]. Fig. 1 depicts the different failure mechanisms of Si-electrode.

The volumetric charge density of alloy anodes (at fully expanded state) is also two to five times greater than the LTO and graphite anodes. Therefore modification of intermetallic space and optimizing morphology are proposed to reduce the observed capacity fade. Adopted methodologies include use of: (a) thin films and amorphous alloys, (b) macro- and nanopowders—nanowires in a brush-type morphology are the most effective in accommodating volumetric strain, (c), different binders and electrolyte formulations—carboxymethyl cellulose (CMC) is the binder of choice for composite Si-anodes, and (d) dispersion in active and inactive matrices. A comparison of various alloy anode materials is presented in Table 2.

4.2.1.3 Conversion anode materials

Insertion reaction requires an electronically conductive active material and crystallographic voids to facilitate Li-ion diffusion. However, Poizot et al.

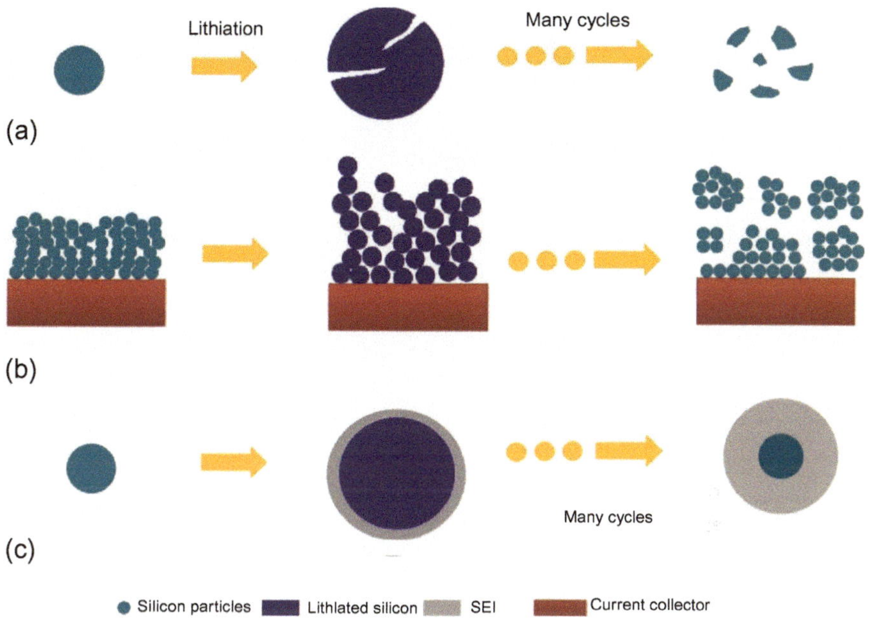

Fig. 1 Failure mechanisms of Si electrode: (a) pulverization, (b) electrode collapse, and (c) continuous breaking and regrowth of the SEI (solid electrolyte interphase) layer [20] (Reprinted with permission).

discovered that the presence of Li-ions can cause various interstitial-free transition metal oxides, for example, Fe_2O_3, CuO, and CoO, which have been classified unsuitable for intercalation reaction, to go through reversible reduction reactions. Some of these metal oxides exhibit reversible capacities up to $1000\,mAhg^{-1}$ [21]. However, the nanostructured reactants must undergo total a chemical and structural transformation as a part of this reaction, represented by Eq. (1) [22]:

$$\mathrm{Nano-MO + 2Li^+ + 2e^- = Nano-M + Li_2O} \tag{1}$$

High-capacity reversible conversion reactions involving as many as four electrons per 3d orbital have been reported for transitional nitrides, fluorides, phosphides, borides, and sulfides as well. Conversion anodes are not considered practical mainly because

1. Development of stable passivation film is unlikely in particular at elevated temperatures because conversion reaction occurs at potentials below the threshold reduction potential of standard Li-ion electrolytes.
2. Energetics of the electrochemical conversion reaction is affected by large hysteresis losses over cyclic discharge process.

Table 2 Comparison of theoretical specific capacity, charge density, volume change, and onset potential of various anode materials (based on information available from [19]).

Materials	Li	C	LTO	Si	Sn	Sb	Al	Mg	Bi
Density (gcm^{-3})	0.53	2.25	3.5	2.33	7.29	6.7	2.7	1.3	9.78
Lithiated phase	Li	LiC_6	$Li_7Ti_5O_{12}$	$Li_{4.4}Si$	$Li_{4.4}Sn$	Li_3Sb	LiAl	Li_3Mg	Li_3Bi
Theoretical specific capacity ($mAhg^{-1}$)	3682	372	175	4200	994	660	993	3350	385
Theoretical charge density ($mAhcm^{-3}$)	2047	837	613	9786	7246	4422	2681	4355	3765
Volume change (%)	100	12	1	320	260	200	96	100	215
Potential vs L (~V)	0	0.05	1.6	0.4	0.6	0.9	0.3	0.1	0.8

4.2.2 Cathodes

Energy density, working voltage, and rate capability of a Li-ion battery are all dependent on thermodynamics and theoretical capacity of cathode materials. The cathode material is, therefore, considered as the key to the development of advanced Li-ion batteries. For the same reason, cathode materials in a Li-ion battery are priced twice as much as anode materials. Cathode materials can also facilitate either intercalation reaction or conversion reaction.

4.2.2.1 Intercalation cathode materials

Intercalation cathodes consist of a solid network of the host material in which guest ions are cyclically inserted and removed. Traditionally, transitional metal compounds, or complex oxides with a spinel, layered or an olivine crystal structure are used as cathode materials. Automotive grade battery cathodes are mainly represented by the oxide family, $LiFePO_4$ being the sole exception. For the advanced Li-ion batteries although, cathode materials with an upper cut-off voltage greater than 4.2 V, for example, Ni-rich layered oxides ($LiNi_{1-x}M_xO_2$ where M = metal ~4.5 V), Li-rich layered oxides ($Li_{1+x}M_{1-x}O_2$ with ~4.7 V), high-voltage spinel oxides ($LiNi_{0.5}Mn_{1.5}O_4$ with ~ 4.8 V), and high-voltage polyanionic compounds (sulfates and phosphates, up to ~ 5.2 V), are recommended [23]. With the help of these materials, the overall contribution of inactive components to battery energy density would be reduced by approximately 20%. Still, cathode materials with practical energy densities close to 800 $Whkg^{-1}$ will be needed to design a battery that can offer 250 $Whkg^{-1}$ [1].

During the electrochemical reaction, Li-ions either take up the vacant tetrahedral/octahedral sites or diffuse into the space between adjacent layers of the host crystal. For charge stability, the same number of electrons as the number of migrating Li-ions must occupy the vacant d-orbitals in the crystal structure of the host transition metal. Electrode potential and the Gibbs free energy charge are determined by electronegativity, ionic radius, valence state, and local environment of the cations. Fig. 2 illustrates the relationship between voltage and the number of electrons in d-orbital of transition metal ions.

Majority of the next-generation cathodes belong to the oxide category because it offers the right balance between energy density, working potential, theoretical capacity, and rate capabilities. Ni-rich NMC oxides with general composition $Li[Ni_{1-x}Co_yMn_z]O_2$, where $x \leq 0.4$ is

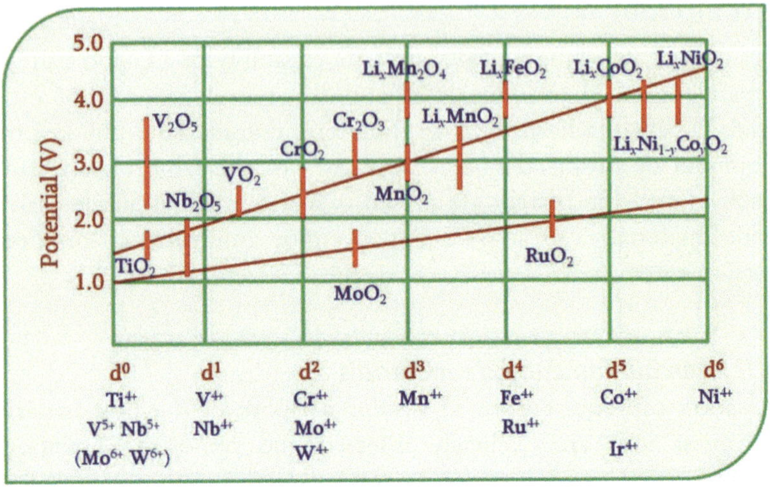

Fig. 2 Voltage range of compounds consisting of transition metal ions. This schematic exhibits the relationship between electrode potential and the number of electrons in *d-type electronic* orbitals of transition metal ions; in general, the potential increases with increasing number of electrons in *d-type electronic* orbitals [24] (Reprinted with permission).

preferred over others by EV sectors because the merits of this class of novel materials increase with increasing Ni content. For example, the reversible capacity of the oxide increases from 163 to over 200 $mAhg^{-1}$ because the composition changes from NMC-111 to NMC-811. Similarly, cathodes using NMC-811 instead of NMC-111 benefit from a higher Li-ion diffusivity (10^{-8}–10^{-9} vs 10^{-10}–10^{-11} cm^2s^{-1}). Electronic conductivity also increases from 5.2×10^{-8} to 2.8×10^{-5} Scm^{-1} upon replacing NMC-111 with NMC-811. Reactivity with electrolyte and structural instability are the issues that need to be resolved before actual implementation.

High specific capacities of high-energy NMCs layered composite materials, general formula: $xLi_2MnO_3(1-x)LiMO_2$ where, M=Ni, Co, or Mn, are attractive to EV industries as well. Li_2MnO_3 serves dual purposes in the composite structure. First, it provides Li^+ to the $LiMO_2$ component and acts as the structural stabilizing member within the composite. Second, it enables a higher degree of lithiation than that usually is achievable with a pure layered oxide. Between 2.0 and 4.4V, $LiMO_2$ is solely responsible for the electrochemical activity. Li_2MnO_3 becomes active as the potential barrier of 4.4V is breached, leading to charge capacities exceeding 250 $mAhg^{-1}$. The process involves dissociation of Li_2MnO_3 into Li_2O and MnO_2. At high C-rates, electrochemical breakdown of Li_2MnO_3 leads to a damaged electrode surface. This causes increased impedance and poor cyclability. In addition, oxygen and lithium loss from the host crystal lead to a large irreversible capacity loss in the Li-rich layered oxides. Coating electrode particles with insulating materials such as MgO, Al_2O_3, $AlPO_4$, and RuO_2 can help to reduce irreversible capacity loss [25]. Generally, nanostructured active material particles provide a large contact area with electrolyte that increases the possibility for side reactions and consequently, electrode destabilization [26].

In terms of the number of potential advantages, polyanionic compounds are rated above transition metal oxides. Majority of the constituent elements of polyanionic compounds are found to be abundant in nature. Consequently, polyanionic compounds tend to be cheaper and eco-friendly than other options. Superior safety properties can also be expected owing to the presence of a strong covalent bond. More importantly, theoretical potential and ionic character of the M—O bond can be easily tuned by selecting the appropriate nontransition metal, for example, B, S, P, and Si [24]. Polyanionic compounds make it possible to design customized or "on-demand" cathode materials. It means that a large pool of available polyanionic compounds with varied crystallographic structures enable to design an energy-dense system and to reduce the working potential if the risk of electrolyte decomposition becomes too high. Selection pool is limited to an extent owing to incompatibility of a high-potential (4.7–5.5 V) group of sulfates and phosphate compounds comprising Co, Ni, and Mn with the current generation of electrolyte. In certain other compounds, an adequate amount of conductive carbon needs to be added to counter high-electronic energy gaps and poor Li^+ diffusion coefficient. Last, the high molecular weight of compounds such as pyrophosphates in comparison to oxides makes it difficult for them to meet the 2025 volumetric energy density targets for batteries [1]. Consequently, they too may be dropped from the selection pool in future.

Some of the technologically mature members of this family are $LiFeBO_3$, $LiVPO_4F$, and $LiMnPO_4$. Deintercalation voltage of $LiFeBO_3$ is nearly 3 V. It offers a theoretical energy density of 660 $Whkg^{-1}$. Besides, excellent cycling stability is ensured by a volume change in the order of less than 2% during cycling. In contrast, inductive effect of fluorine causes $LiVPO_4F$ to operate at a potential close to 4.26 V. However, the energy density of $LiVPO_4F$ is quite similar to that of $LiFeBO_3$ [650 $Whkg^{-1}$ (or 2000 Whl^{-1}) vs 660 $Whkg^{-1}$]. It also demonstrates fast Li-ion diffusion rates and a good long-term reversible capacity retainability. $LiMnPO_4$ too provides energy storage capacity above 600 $Whkg^{-1}$. A few other manganese-based compounds, for example, Li_2MnSiO_4 and $LiMnPO_4(OH)$, operate at 4.5 V. They provide gravimetric and volumetric capacities close to $1300\,Whkg^{-1}$ and $4000\,Whl^{-1}$, respectively. Electron transfer process initiates multistep electrode reactions. In most cases, only one electron transfer is detected because it occurs at a potential that is quite close to the breakdown potential of a standard electrolyte solution. However, two distinct electron transfer steps are encountered in certain molecules, mainly with two identical electroactive groups. Two-electron transfer process is likely responsible for their impressive performance. Performance can be further improved by modifying transport properties, where low Li-ion diffusion coefficients and low electronic conductivities are reported for silicates ($5 \times 10^{-16}\,Scm^{-1}$ at 25°C).

4.2.2.2 Conversion cathode materials

Intercalation reactions typically involve one electron transfer for each metal atom. In contrast, another category of cathodic materials, known as conversion materials, which support multielectron reactions and offer theoretical gravimetric densities up to $700\,mAhg^{-1}$, also exist. Reaction mechanism is quite similar to anodic conversion reactions in which metal atoms are reduced to their metallic states. The reaction product is a Li_nX matrix containing metal nanoparticles in a dispersed state, where X is an anion, for example, F, N, P, S, etc., and n is the oxidation state of X in the conversion compound M_aX_b.

Conversion compounds are known for their low molar weight, low cost, and high intrinsic thermal stability. It should, however, be noted that the operating potential of the majority of the conversion compounds is lower than the operating potential of oxides and polyanionic compounds. Furthermore, it strongly depends on the reaction potential of the negative ion forming the compound. They are also characterized by low ionic conductivities, large hysteresis, and poor cycling stability caused by particle delamination owing to large volume changes during lithiation and delithiation processes. Some of the compounds, for example, $CuCl_2$, are soluble in organic solvents or exhibit polysulphide dissolution. Armstrong et al. pointed out that consistent particle morphology and systematic material preparation is the key to the reliable performance of conversion cathodes because conversion materials prepared through different synthetic methods demonstrate significantly different behaviors [27].

Material loading in the present generation EV batteries is between 15 and 25 $mg cm^{-2}$. However, most of the novel cathode materials are tested with a loading as low as 1 $mg cm^{-2}$ [1]. Rigorous testing is, therefore, needed before any of the novel materials can be applied in traction batteries.

4.2.3 Electrolytes

A polar aprotic Li salt, typically lithium hexafluorophosphate ($LiPF_6$) in a binary nonaqueous solvent mixture, including either ethyl carbonate (EC) or ethyl methyl carbonate (EMC), diethyl carbonate (DEC) or dimethyl carbonate (DMC), is used as standard electrolyte in currently available commercial Li-ion batteries. High anodic stability, high ionic conductivity even at the temperature as low as −15°C and high solubility of $LiPF_6$ in all alkyl carbonates are the key reasons for the preferred selection. Usually, anodic stability and oxidation potential of electrolytes increase as the concentration of atoms in high oxidation states increases. However, the presence of Li-ions alters the reaction thermodynamics completely. Salt anions of the type MX_y and polar aprotic solutions are reduced to insoluble ionic Li compounds. Reduction compounds precipitate over carbon electrodes and nonactive metal electrodes to form a thin surface film at the electrode and electrolyte interphase. This thin surface film allows Li-ion transfer but starts blocking electron flow after attaining a certain thickness. The observed passivating effect is related to the apparent stability of commercial electrodes. The DMC and EC solutions are known to form highly stable passivating surface film over graphite anodes upon reduction.

The viscosity of these solutions is relatively high. Consequently, wetting process for composite electrodes and separators becomes tricky. In addition, alternative electrolyte solutions are required for low temperature applications because the ionic conductivity of the EC/DMC solutions decreases remarkably below −15°C. Decomposition of $LiPF_6$ to PF_5 and LiF at elevated temperatures is also an issue. More importantly, the standard electrolyte solvents pose significant safety challenge during handling, usage, and events such as thermal runaway owing to their high flammability and high toxicity. Compatible electrolytes are crucial for enhanced performance, cycle life, and safety of advanced Li-ion batteries.

Organic solvents such as cyclic phosphates and halogenated cyclic carbonates exhibit better fire retardancy than EC/DMC solution. Water sensitivity, high viscosity, and poor stability are some issues that need to be addressed [28–30]. Use of aqueous electrolytes, developed by dissolving salts such as $LiNO_3$, Li_2SO_4, and LiTFSI into water such that the molality exceeds 20 M, is a promising approach to increase fire retardancy of Li-ion batteries [31]. Nonvolatile electrolytes based on ionic liquids are also being investigated for the next-generation batteries. Ionic liquids demonstrate nonflammability. Nonetheless, their practical applications are restricted owing to their high costs and incompatibility with carbonaceous anodes. Formation of SEI film on graph cyclic phosphate is difficult in Li-ion batteries using ionic liquids. Li-ion migration is also reduced as cations of ionic liquids compete with Li-ions and occupy the intercalation sites in graphite layers [32–34]. It must be noted that average working voltages of organic electrolyte solvents lie between 1.5 and 4.5 V. Corresponding energy density values are in the range of 150–300 $Whkg^{-1}$. In contrast, solid-state electrolytes are more stable, safe, and compatible with new-age high-voltage cathode materials. It can achieve energy densities above 500 $Whkg^{-1}$ [35].

4.3 Battery manufacturing methods

Significant variability is introduced in battery cells during the manufacturing process itself. The sequence of operation, for example, mixing and dispersion steps and the process parameters such as drying time and temperature have a profound effect on the performance of manufactured cells. This section provides a concise description of the conventional battery manufacturing method and introduces some advanced techniques developed in the recent years.

4.3.1 Classical method: Wet slurry casting

Commercial batteries are manufactured using a wet slurry casting method. It is a complex technique involving several sequential operations. A simple illustration showing the necessary manufacturing steps is presented in Fig. 3.

The process begins with mixing the active material, the conductive material, and the binder material continuously. Mixing time and temperature can vary between 30 min and 5 h and from 20°C to 40°C, respectively. Subsequently, the dry mixture is homogeneously dispersed in a solvent under vacuum. Vacuum is used to avoid gas inclusions. Different uniform and viscous pastes, commonly known as slurries, are prepared for the anode and the cathode electrodes. Solvent is added for adjusting the viscosity of the slurry, whereas the binders are inert compounds responsible for maintaining the mechanical integrity of the

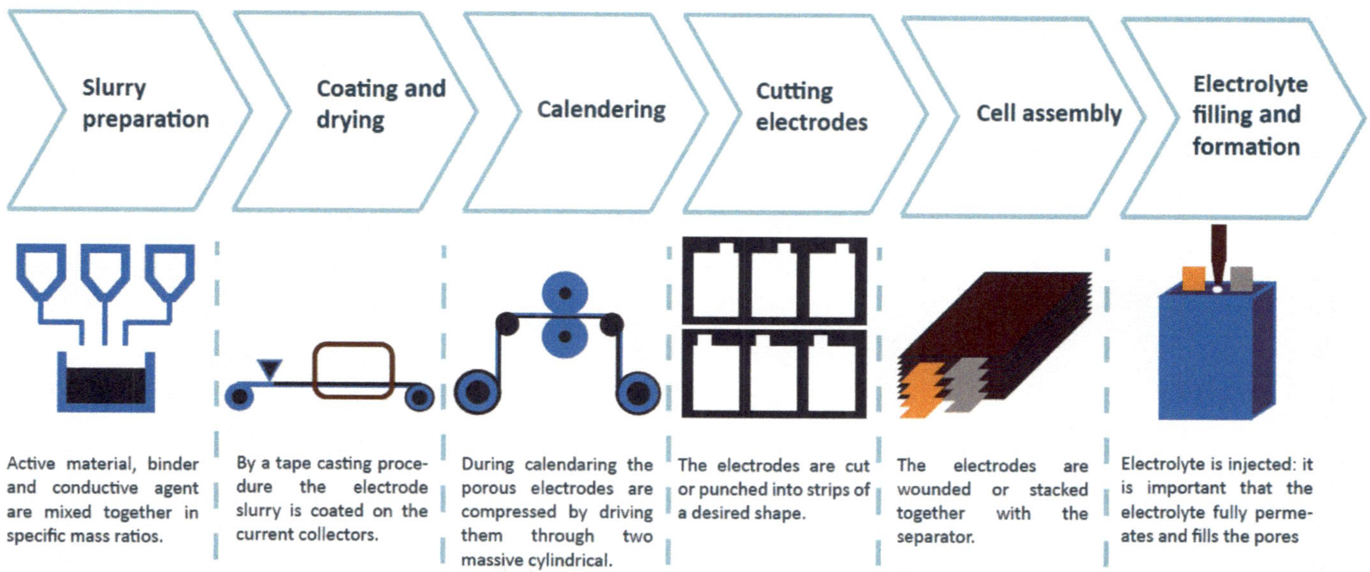

Fig. 3 Steps in wet slurry-based battery manufacturing process (Available under CC BY 4.0 license from [36]).

slurry. They provide adhesive and cohesive forces needed for holding the dried slurry together onto the electrode film. For casting the cathode, a slurry with a viscosity in the range of 1200–500 cPa s at a shear rate of 3–8 s^{-1} is preferred. On the other hand, a slurry with a viscosity between 1400 and 600 cPa s at a shear rate of 3–8 s^{-1} is used for casting the anodes.

The mixing and forming of a slurry electrode include: (i) raw material conveyance and/or feeding, (ii) mixing, (iii) mixed slurry conveyance, (iv) dispensing and/or extruding, and (v) forming. The raw material is fed in batches, but the conveyance and/or feeding can be natural or forced. In natural feeding, the raw material is supplied to the mixer at a rate that allows the mixer to accept raw material without any external force. On the other hand, a piston mechanism or a screw-based "side-stuffer" provides the required pressure in forced feeding conditions. The slurry is coated onto the surface of a current collector—aluminum foil for the cathode and copper foil in the case of the anode electrode. Standard industry practice is to use a blade coating or a slot die machine for uniformly spreading wet electrode slurry onto a moving current collector foil. Typically, the blade or the die head is fixed 5–200 μm above the foil. Larger gaps between the blade and the current collector foil tend to create unevenly coated surfaces upon drying. In contrast, smaller distances lead to the application of imperfect films. Practical issues such as loss of coating integrity, adhesion, and excessive flaking also limit the maximum coating thickness on single-side to approximately 100 μm [37].

Coated electrodes are dried and then calendered to control the thickness and to increase the energy density of the final output. The calendered electrodes are cut and costacked (for a pouch cell assembly) or cowound (for cylindrical formats) with a thin intervening separator membrane to fabricate battery windings. After packaging the windings in an appropriate metallic or plastic casing, the liquid electrolyte is injected in it under a controlled environment. The calendering process, however, reduces the porosity of the coated electrode layer by 65%–70%. This has a direct impact on pore-size distribution and electrode wettability [38]. Consequently, a period of 2–3 days with vacuum assist (down to 25 mm Hg absolute) for up to 24 h is required to achieve adequate electrode wetting, crucial for stable SEI layer formation. Still, a large portion of the smallest pores remain unwetted, which has an adverse effect on cyclability of conventionally manufactured battery cells [37]. The final thickness of the dried electrode depends on the concentration, morphology, viscosity, and surface chemistry of the ink or slurry.

4.3.2 Advanced manufacturing methods

Despite their high costs, classical wet slurry coating-based battery manufacturing processes offer neither quick production turnaround nor high quality control. Moreover, additives used for dissolving the binder material are toxic solvents, which react with electrode materials and affect their physicochemical properties. Therefore alternative battery manufacturing methods are being pursued.

4.3.2.1 Solvent-free dry electrode coating

US Patent 8213156 discloses a solvent-free dry electrode manufacturing method. The term "dry" refers to no use of additive solvents in the manufacturing process, meaning that liquid-free status is maintained right until the final electrolyte-impregnating step. This unique method eliminates toxic solvents used for dissolving the binder from the production chain. Furthermore, it enables manufacturing electrodes with active material layer thicknesses up to 1 mm, making it an eco-friendly and an economically attractive manufacturing method at the same time. The process includes three steps:

1. Dry powder mixing—a mixture of activated carbon particles, conductive carbon particles, and fibrillizable binder is blended for 1–10 min in a V-blender with a high-intensity mixing bar to form a homogeneous dry mixture. Blending time varies depending on materials, particle size, batch size, density, etc.
2. Film formation—binder particles in the dry mixture are fibrillized through the application of high shear forces through a jet mill or through application of high-pressure gas (pressure $\geq$ 60 psi) having a dew point between −40°F and −20°F and water content less than 20 ppm. Fibrillization physically stretches the binder particles producing thin web-like fibrils. Eventually, the fibrils form a lattice or matrix that entraps, binds, and supports the resulting composition of the dry powder mixture. Subsequently, the dry powder is calendered by passing through a roll mill. A self-supporting dry film of thickness ranging between 100 and 250 μm is formed after a single pass through the roll mill.
3. Laminating film to current collector—once a self-supporting dry film of target specification is prepared, it is calendered on to a current collector foil. Dry film density produced can range between 0.5 and 0.7 gcm^{-2}. An intermediate adhesive layer could be used between the current collector and the dry film, or the dry film could be laminated on to pretreated substrate directly.

Compressed feed air transfers the mixture of active materials, conductive carbon, and binder material in appropriate ratio to a ceramic-lined grinding chamber of the jet mill. The ceramic lining helps to maintain purity of the resultant jet-milled dry mixture by minimizing abrasive wear of the chamber's internal walls. Multiple nozzles arranged circumferentially on the chamber walls inject compressed grind air inside the chamber. Grind air accelerates the dry mixture resulting in mainly particle-to-particle collisions. Particle collisions cause fibrillization of the dry binder and may affect size reduction of the active material/carbon aggregates and agglomerates.

Fibrillized binder particles create an interconnected network of point contacts between the active material particles and the conductive carbon particles. Dry binder network is less obtrusive and enables intimate electronic contact and high ionic conductivity. Consequently, electrodes manufactured through dry coating technology exhibit lower charge transfer impedance and lower particle-to-particle contact resistance, thus facilitating high rate performance in high-energy density Li-ion batteries. Colliding dry mixture particles spiral toward the chamber center and exit through a discharge hose. Discharge hose directs the exiting dry mixture to a dust collector where the mixture is separated into air and dry fibrillized active material/carbon/binder mixture. Air is filtered, as many times as required, and discharged in atmosphere, whereas the dry mixture is collected in a container.

Fig. 4 shows an image of freestanding electrode films manufactured through this process. Physicochemical properties of the manufactured electrode depend upon fibrillization grade of the fibrillizable binder material and consistency and homogeneity of the resultant dry mixture. For example, the tensile strength of the fabricated electrode is controlled by both the degree of fibrillization of the binder and the consistency of the matrix created by the binder fibrils within the dry material used for fabricating the electrode. On the other hand, the internal resistance of the electrode is a function of bulk resistivity of the dry film, which in turn depends on the homogeneity of the fibrillized mix. Accordingly, better dispersal of conductive fillers in the fibrillized mix implies higher homogeneity and lower internal resistance for the dry coated electrode. However, it has been observed that voids appear between the active material/carbon particles of the dry mixture exiting the dust collector as they begin to settle and form clusters under the influence of gravitational forces. As a result, the dry mixture can retain its homogeneity and

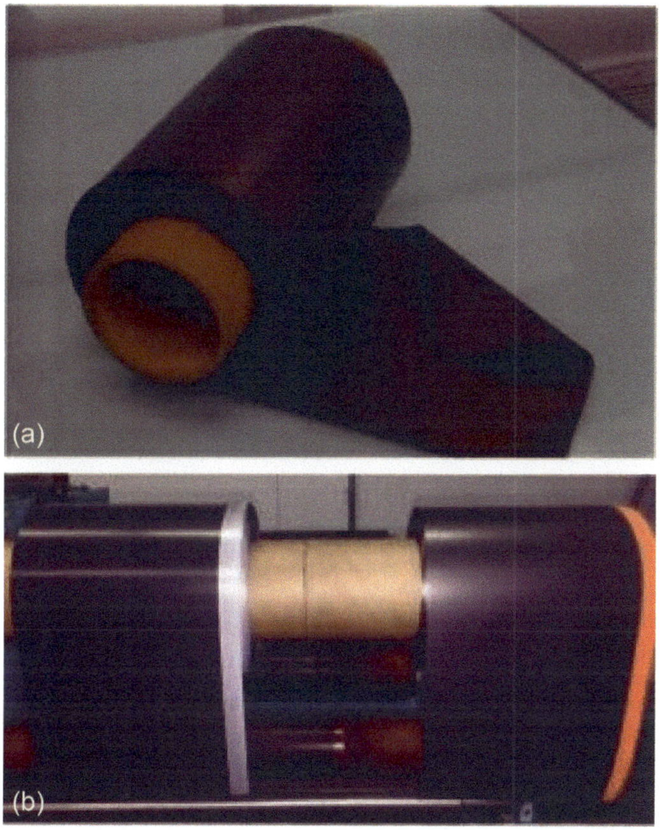

Fig. 4 Dry battery electrodes (a) a roll of freestanding dry electrode film fabricated using roll-to-roll equipment, (b) double-sided dry NMC electrode roll (left) and graphite roll (right) laminated onto current collector, ready for cell assembly [39] (Reprinted with permission).

consistency for a limited period of approximately 10 min only. Hence, it is imperative that either the homogeneous mixture is processed within this small window, or it should be kept adequately aerated to ensure material consistency and tight quality control in downstream processes.

Other factors that influence tensile strength and resistivity of the dry film manufactured using the jet-milled dry mixture are (a) feed rate, (b) feed air or inject air pressure, and (c) grind air pressure. A set of experiments was designed by the inventors to investigate the effect of these three parameters on the qualities mentioned earlier. Table 3 presents resistivity measurements for the jet-milled dry blend of particles provided by dry fibrillization step in a range of conditions.

Table 3 Resistance of the jet-milled dry mixture under different conditions.

Feed rate (kgh^{-1})	Feed pressure (psi)	Grind pressure (psi)	Dry resistance (Ohm)
3	60	85	0.267
3	70	85	0.229
3	60	110	0.221
3	70	110	0.212
9.6	60	85	0.233
9.6	70	85	0.208
9.6	60	110	0.241
9.6	70	110	0.256

Table 4 Average tensile strength of the dry film in width-direction and in length-direction measured at different feed rates, feed pressure, and grind pressures for samples of different thicknesses.

Feed rate (kgh^{-1})	Feed pressure (psi)	Grind pressure (psi)	Sample thickness (mm)	Tensile strength in (g)	
				Width	Length
3	60	85	6.1	63.00	123.00
3	70	85	5.5	66.00	146.00
3	60	110	6.2	77.00	166.00
3	70	110	6.1	59.00	108.00
9.6	60	85	6.0	58.00	132.00
9.6	70	85	5.8	70.00	145.00
9.6	60	110	6.0	61.00	135.00
9.6	70	110	6.2	63.00	141.00

Subsequently, the dry mixture was processed to make dry electrode films, which were kept in vacuum at 160°C for 3 h to remove traces of moisture. Five strips of standard width were prepared as samples for each experiment, and a reliable estimate was derived by averaging the measured values. Table 4 presents the averaged values of the tensile strength measured along the width and along the length of the samples.

Table 5 Comparison of the electrodes made by wet and dry processes (based on information available in [41]).

Electrodes	Thickness (μm)	Mass loading ($mgcm^{-2}$)	Porosity (%)	Packing density (gcm^{-3})	Discharge capacity ($mAhg^{-1}$)
Wet slurry-based (low loading)	32.6	7.65	41	2.5	157
Wet slurry-based (high loading)	52	14.27	35	2.7	156
Dry powder-coated	40.5	10.07	31	2.4	155

It is clear from the tabulated data that the effect of inject pressure or the feed air pressure on the resistance of the dry mixture dominates the effect of grind air pressure and the effect of the feed rate. The resistivity of the dry electrode decreases as the feed air pressure is increased. In contrast, the tensile strength of the electrode increases with the increasing inject pressure. It can, therefore, be implied that increasing inject pressure has a largely positive effect on physicochemical properties of dry film-coated electrodes [40]. Al-Shroofy et al. compared the conventional wet slurry-based electrodes (with low and high loading) to dry powder-coated electrode. Table 5 presents some findings from their analysis. The dry electrodes exhibited high specific discharge capacity and good capacity retention in comparison to the electrodes fabricated using the wet slurry [41].

Dry electrode manufacturing technique has several advantages over the classical wet slurry casting method. Some of the key benefits are

- It reduces the additive-based impurities and eliminates volatile waste products from the system.
- A larger pool of binder materials is available for selection because they do not need to be dissolvable unlike the binders used in wet slurry-based manufacturing.
- 10%–20% lower operating costs owing to reduced dwell/drying times.
- Solvent recovery and recycling systems are made obsolete, which leads to a need for lower capital investments.
- Highly dense electrode microstructures provide improved energy and power density.
- Cycle life enhanced by a factor of two owing to superior adhesion and cohesion properties in the presence of electrolyte at high temperatures.

The solvent-free dry electrode coating method is a promising technique with the potential to become the go-to manufacturing process for the next-generation batteries. Thus Tesla, Inc. recently acquired Maxwell Technologies, Inc., an American developer and manufacturer of battery cells with dry coated electrodes in an all stock exchange offer [42,43].

4.3.2.2 Semisolid battery

In the classical approach, thin metallic foils coated with a wet slurry are dried in a series of hot temperature chambers before further processing. It takes approximately 22 h for the foils to completely dry. Duration of drying time is a matter of serious concern in the wet slurry casting method. US patent 10566603 presents a fundamentally different and a simplified battery manufacturing process that requires one-fifth of the time needed to create a battery cell conventionally. The disclosed method proposes a semisolid battery in which the electrode coating is kept in a liquid state through all the production steps instead of waiting for it to dry before proceeding to subsequent steps. The drying process is completely removed as a production step in this design. In addition, the proposed semisolid battery does not require any binding material and calendering and solvent recovery steps.

The semisolid battery consists of two compartments, anodic and cathodic, separated by an ion-permissible membrane. Anodic compartment refers to the space between the anode current collector (copper) and the separator. Similarly, the cathodic compartment is the space between the cathode current collector (aluminum) and the separator. Semisolid gels for the anodic compartment and the cathodic compartment are prepared by mixing electrochemically active material and electronically conductive particles in appropriate ratios and cosuspending them in an electrolyte solution. The semisolid mixtures can flow under the effect of gravitational field or with the aid of mechanical vibrations or upon application of a pressure differential or an acceleration field that exerts a force on the gel.

The battery is manufactured by transferring the anodic and the cathodic semisolid suspension gels in the anode compartment and the cathode compartment, respectively, and sealing the compartments thereafter. The compartments can be filled through injection nozzles or by using the ports made along the edges of the cell. Mechanical vibration, gravitational force, sonic agitation, or shearing forces are used to assist the filling process. Vacuum assist creates suction in the electrode compartment to ensure complete filling. Mechanical/vacuum assistance augments flow-inducing forces, promotes suspension stability, decreases viscosity, and enables maintaining an activated carbon network. "Activated carbon network" is a qualitative state referring to morphology and arrangement of individual carbon particles within the electrode. It is characterized by enhanced electrical contact and conductivity among carbon particles present in the electrode. Furthermore, concurrent filling of both the electrode compartments is preferred to balance deflection forces acting on the separator and to minimize its deformation.

The energy density of the semisolid battery is proportional to the packing density of the semisolid suspension. A flowable semisolid suspension with high packing density is achieved by using a polydisperse size distribution of active material particles in slurry preparation. Distribution should be such that the size of the largest particles available in 5% by volume of the suspension gel is bigger than the size of the finest particles available in 5% by volume of the suspension gel by at least a factor of five. Spherical particle morphologies assist in achieving high particle densities. Simultaneously, semisolid suspensions consisting of spherical particles have low viscosity and tend to flow well. In contrast, semisolid suspensions containing equiaxed particles have good flowability but low packing density. Typically, particles with large size distribution are used to prepare the slurry/semisolid suspension. It improves the material flow during cell loading. Smaller particles fit in the interstitial spaces of the larger particles and contribute toward the increased solid volume fraction and high packing density of the loaded cell. Moreover, surfactants and dispersants such as stearic acid are added to the mixture for increasing particle packing density along with flow ability while decreasing viscosity. Last, charge transfer properties of the semisolid suspension are improved by adding redox mediators based on Fe^{2+} or V^{2+}, V^{3+}, or V^{4+}.

Li salt concentration in the semisolid suspension is increased after filling the electrode compartments to densify the particle suspension. The resulting mixture has an electronic conductivity between approximately 10^{-6} and 10^{-3} Scm^{-1}. Electrical double layer, responsible for causing repulsion between particles, collapses as salt concentration in the semisolid suspension increases. In the absence of repulsive forces, particles begin to aggregate, which in turn leads to the densification of the semisolid suspension. This allows high-density electrodes to be manufactured from a low viscosity suspension by pumping it or injecting it into the net-shaped electrode compartments.

A significant percentage of inactive materials is present inside a cell's casing. These materials are not only wasteful and expensive but add dead weight to the ESS as well. The relative contribution of the inactive components per unit volume of the cell decreases as the cell thickness and the material loading are increased [5,44]. Nevertheless, the rate capabilities of conventionally designed electrodes suffer as their thickness increases. Reducing tortuosity is the key to creating thick electrodes with fast charge-discharge characteristics. A semisolid suspension provides more line-of-sight paths for Li-ion transfer in comparison to a classically constructed electrode that offers a tortuous path in a network of inactive material. The semisolid battery design further reduces the inactive material percentage in a cell by more than 80%. Simultaneously, it allows for achieving a fivefold increase in the active material coating thickness. Moreover, the semisolid electrode formulation makes the new batteries flexible and resilient in contrast to the classical electrodes that are brittle and crack under stress. Foldability of the semisolid batteries allows them to clear the nail penetration test without failing, thus increasing their durability and safety [45].

4.3.2.3 3D-printed batteries

The three-dimensional (3D) battery architectures are another possible means of designing thick, high-energy density electrodes. They decouple the energy density and power density trade-off generally associated with the design of thick battery electrodes by increasing the ratio of effective surface area to volume, thereby shortening the Li-ion diffusion pathways. Additive manufacturing (AM) technology, also known as 3D-printing technique, is an umbrella term for shaping principles that build 3D profiles through successive addition of material [24]. Innovative aspect or specialty of 3D-printing technology is that each layer can be manufactured from a different material and can have different process parameters and geometry [25]. Besides, the AM processes are theoretically more flexible and economical in comparison to the conventional manufacturing methods.

According to the American Society for Testing and Materials (ASTM), the AM techniques are classified into seven categories, governed by the ASTM Committee F42 on Additive Manufacturing. They differ mainly based on the form of the raw material used (powder, liquid, filament), heat or light energy source (thermal, laser, electron beam, plasma arc), feed systems and build chamber characteristics, and the number of print axes. Also, some of the techniques require additional postprocessing. The categories are as follows:

- Powder bed fusion uses thermal energy to fuse particles of a powder by melting it. The energy source is a beam of electrons, a laser, or a heated print head. Techniques include direct metal laser melting (DMLM), direct metal laser sintering (DMLS), electron beam melting (EBM), selective heat sintering (SHS), among others.
- Vat photopolymerization uses a light source instead of heat to provide the activation energy. Another significant difference is that raw material is used in liquid form rather than as a powder or as filaments. Examples include stereolithography (SLA) and continuous direct light processing (CDLP) methods.
- Material jetting is one of the more precise AM techniques in which tiny droplets are dispensed through nozzles layer by layer using a drop-on-demand technology. A second series of nozzles dispense a dissolvable polymer for providing support to the object during printing. The supporting material is washed away after the printing is complete. Structures as thin as 15 μm can be printed with this method.
- Material extrusion is a fast and relatively inexpensive method wherein thermoplastic filaments are extruded in the horizontal *x*- and *y*-directions from heated nozzles to make 3D parts. Anisotropy or strength in the vertical, that is, *z*-direction can be an issue for extruded components.
- Directed energy deposition is a multiaxis printing technique. In this, electron beam, plasma arc, or a laser is focused on selected wire filaments or powdered metals to melt and fuse them inside a vacuum chamber. Methods include rapid plasma deposition (RPD), laser engineered net shape (LENS), and electron beam additive manufacturing (EBAM). Minimum thickness of an individual layer is 20 μm.
- Binder jetting is similar to material jetting except that a binding agent is used in addition to powdered raw material to increase the mechanical properties of the functional component. Postprocessing is needed for parts produced through binder jetting.
- Sheet lamination involves bonding ultrathin sheets of different solid materials such as paper and metal by using adhesives. Techniques like laminated object manufacturing (LOM), paper lamination technology (PLT), and ultrasonic additive manufacturing (UAM) are grouped under this category.

Table 6 presents a qualitative comparison of selected 3D-printing techniques from the aforementioned categories. Among them, inkjet printing (IJP) and

Table 6 Qualitative comparison of selected 3D-printing techniques.

3D-printing technique	Ink characteristics	Suitable materials	Printing resolution	Multimaterial capability	Advantages	Disadvantages
TAE	Newtonian liquids	Metals, polymers, oxides, and hydroxides	50146nm–10146μm	Good	– Less demanding setup – Low cost and versatile – Enables printing highly ordered macroporous structures – Enables complete filing of interstitial voids from bottom to top layers	– Limited range of printer compatible materials – Printed components affected by weak mechanical properties
SLA	Photosensitive resins	Photopolymers	0.25–10 μm	Challenging	– High resolution – High surface finish	– Highly material-specific
DIW	High viscosity non-Newtonian paste	Metals, plastic, ceramics, and organic compounds	1 μm	Good	– Can be applied with a variety of materials – Simple operation – Affordable	– High scrap generation – Printed parts may have weak mechanical properties

AJP	Aerosols	Nanomaterials, dielectric materials	10 μm	Very good	– Ink viscosity does not need to be optimized – High efficiency	– Costly equipment – Unstable print quality
IJP	Low viscosity Newtonian liquid	Carbon-based and protein materials, conductive polymers, sol-gel, and metals	20 μm	Very good	– Can print large objects – Low cost	– Needs careful ink rheology optimization – Print head requires regular maintenance – Low print speeds
FDM	PLA or ABS filaments doped with active materials	Thermoplastics	50–200 μm	Possible	– Can print large parts – Chemical postprocessing is not needed – High speed and simple operation	– Low surface finish – Limited *z*-axis resolution

AJP, aerosol jet printing; *FDM*, fused deposition modeling; *SLA*, stereolithography; *TAE*, template-assisted electrodeposition.

direct ink writing (DIW) are most commonly applied for printing batteries. They allow using active materials' dispersion in a solvent directly as printing ink. In contrast, other methods require special material engineering and processing steps for preparing compatible inks.

4.3.2.3.1 Inkjet printing

IJP is a droplet-based, direct deposition technique for liquid phase materials [46]. Complex patterns can be easily created as high resolution parts using IJP technique. It has also illustrated excellent multimaterial printing capability. Therefore it has been applied in a variety of fields ranging from biomedical to electrical [46]. In principle, it can be either a material (polymer) jetting process or a binder jetting process. In the former, a photosensitive polymer ejects from the inkjet on to the substrate, whereas in the latter, a binder is selectively deposited on active material powder bed to form the part.

Because the rheology and formulation of the ink are quite similar to that of the slurry used in conventional die casting method, IJP has found wide applicability in 3D-printing of batteries. Notwithstanding, the ink used for IJP should be in a dilute form, that is, not too high active material concentration, without any agglomerate to prevent clogging of the nozzles. It is a common understanding that ink characteristics like density (ρ), viscosity (μ), and surface tension (σ) must be customized in accordance to the fixed nozzle diameter (d) for the IJP process to be successful. Specifically, the inverse Ohnesorge number, Z, defined by Eq. (2), must be between 1 and 10 for the ink to be able to form stable inkjet droplets [47]:

$$Z = \frac{d\sigma\rho}{\mu} \tag{2}$$

Lawes et al. utilized a desktop inkjet printer for fabricating Li-ion batteries with thin-film Si anodes. The ink was formulated by mixing Si nanoparticles (active material) with a polymer binder and carbon black (conducting agent) in water uniformly through sonication. The printed batteries showcased discharge capacity greater than 1700 mAhg^{-1} for 100 cycles. Besides, a stable cycling performance was measured during 1000 cycles of limited discharge equaling 1000 mAhg^{-1}. They also recommended poly(3,4-ethylenedioxythiophene): polystyrene sulfonate (PEDOT: PSS) over other binders for manufacturing battery electrodes [48]. Hu et al. used a slurry of $LiMn_{0.21}Fe_{0.79}PO_4$ (LMFP) nanocrystals for fabricating Li-ion battery cathodes through IJP process. NMP was used as a solvent for making the slurry. Carbon black polyvinylidene fluoride (PVDF) were added in the slurry mix as the conducting agent and the binding agent, respectively. Cathodes of a well-controlled height and width were printed on an aluminum foil. Half cells were prepared by assembling the printed cathodes with a counter electrode (lithium foil) and filling the cell with electrolyte. Gravimetric discharge capacity measured at rates between 1C and 100C for the printed cells was noted to be superior to that of the conventionally made cells under similar conditions (161.36–108.45 vs 141–70 $mAhg^{-1}$). Similarly, the specific capacities measured after 1000 cycles at 10C for the printed cell and the conventional cell were 150.21 and 103.38 $mAhg^{-1}$, respectively. The respective values corresponding to a 20C discharge rate were 140.67 and 90.64 $mAhg^{-1}$ [49]. Milroy et al. manufactured less than 10 μm thick lithium-sulfur cathodes for integrated nanomanufacturing using single-walled CNTs infused with straight chain sulfur through IJP process. The printed electrodes exhibited a capacity of approximately 700 $mAhg^{-1}$ at 0.5C rate after 100 cycles [50]. IJP is a powerful tool for precision patterning. It offers a high material utilization capability along with a rich material selection. There is a significant potential to engineer energy-dense electrode structures with high rate capability in a very accurate manner using this technique.

4.3.2.3.2 Direct ink writing

Direct ink writing (DIW) is also a direct material deposition technique. However, instead of being in a liquid phase, the ink should be in a paste-like form exhibiting shear thinning behavior. The ink must also possess a high storage modulus and a high yield stress that enables a distortion free connection between continuous filaments and shape retention of extruded lines. It is a versatile process. At the same time, it is comparatively less challenging owing to the relatively smaller risk of nozzle clogging than in the IJP process. More importantly, it is possible to achieve a high material loading with DIW process. As a result, it has been selected by several teams for battery manufacturing purposes [51–54].

For example, to fabricate a battery with thick electrodes, all its components, that is, both the electrodes, separator and packaging were printed using the DIW method by members of the Lewis group. Four different types of inks were formulated with specific viscoelastic and shear thinning response to match typical characteristics of the four components. Electrodes of thickness ranging between 50 μm and 1 mm were printed. LFP and LTO were used as the active material for cathodes and anodes, respectively. The ultrathick electrodes manufactured through this process demonstrated higher areal energy density than the thin electrodes. Another equally important observation was that the areal power densities were also comparable [55]. Sun et al. used the DIW technique to print a 3D interdigitated microbattery with LFP cathode and LTO anode. The inks were designed by mixing the active material with additives—glycerol and ethylene glycol and a cellulose-based viscosifier in deionized water. Multilayer electrodes were extruded on Au current collectors, which were then dried in an inert gas atmosphere at 600°C to remove the organic additives. The printed microbattery demonstrated storage capacity of $1.2\,mAhcm^{-2}$ at 0.5C. Moreover, power and areal energy densities of $2.7\,mWcm^{-2}$ and $9.7\,Jcm^{-2}$, respectively, were measured [56]. Fu et al. also employed the DIW process to prepare an all component 3D-printed battery. GO-based composite inks were used for the electrodes, whereas a gel polymer composite ink served the purpose of the electrolyte and the separator. Furthermore, the inks were aqueous-based. Consequently, there was no need for sintering at high temperature for removing the organic additives [57].

4.3.2.4 Cell-to-pack technology

Rechargeable energy storage systems or battery packs are created by combining multiple parts such as individual battery cells, current transmission system, that is, electrical connectors and associated wiring harness, battery management system, various sensors, safety features and mechanisms, and structural support. Individual battery cells can be either small format (cylindrical cells) or large format pouch cells and prismatic cells. Employing small format cells in pack design has many advantages. For example, manufacturing processes for electrodes of short length are easy to control. Therefore they are typically defect-free. As a result, manufacturing yield is also greater than those for large format cells. Moreover, their structural design makes them inherently safe as it is relatively easy to control any volumetric expansion during (dis)charge cycles and thermal events. However, packaging becomes difficult as correct alignment and proper electrical connection between many small format cells need to be ensured. Furthermore, the auxiliary component count or the percent of inactive components increases with the number of individual cells in the battery pack, which adds to the high manufacturing cost.

The US patent US20190312251 discloses a concept that leverages the advantages offered by the cylindrical cells while making manufacturing of battery packs simpler, less complicated, and cost-effective than the prior

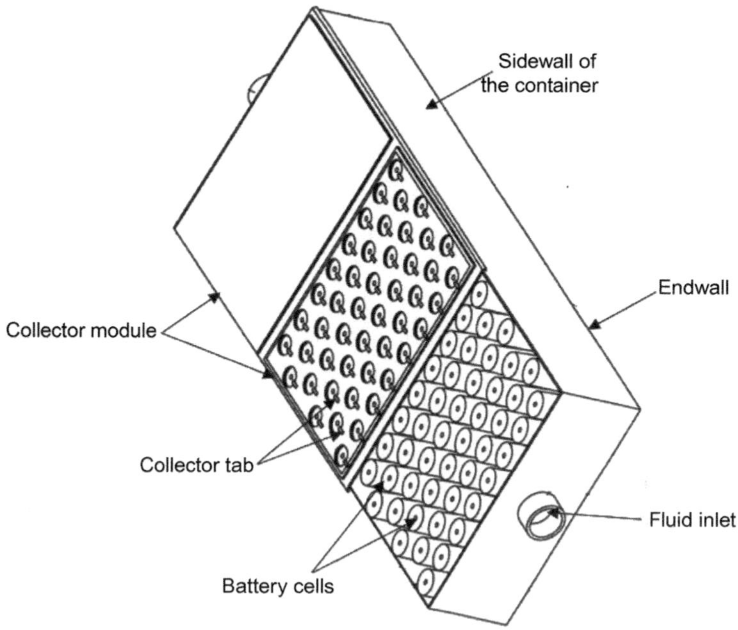

Fig. 5 Illustration of battery cells connected through collector plate and placed in a container that has an inlet and outlet for cooling fluid.

art. For example, the cells are configured with a common tap gap manufactured from a single part using a die or a stamping tool. A common cap reduced cost and complexity.

The disclosed design includes a container with multiple sidewalls adjoining one another and bound by an end wall. Another end of the container is purposefully left open as seen in Fig. 5. Through this opening, battery cells are positioned inside the container such that their longitudinal axes are parallel to one another. Upright orientation is preferable for cells. The container includes two collector modules or collector plates that are designed by stacking or overlaying a dielectric matrix on top of metal plates or printed circuit boards through techniques such as welding, bonding, or adhesion. Multiple tabs are then incorporated within the dielectric matrix at predesignated locations corresponding to the position of battery cells in the container such that they partly extend in an outward direction. The embedded regions of the tabs are connected to one another internally, that is, within the volume of the dielectric matrix, while the extended portions are used for forming electrical connections between different cells. The dielectric or an insulation layer enables placement of both the collector plates on one side, for example, top of the battery cells without any risk of short circuit.

Tabs from the first plate connect cathodes of the battery cells located in the container, whereas the tabs in the second collector plate are used for electrically connecting anodes from the respective cells to make one battery block or system. In addition, the collector plates also feature an overhang portion that is useful for interconnecting battery systems. The current collectors have in-built sense circuits, fuses, and resistors needed to control the battery system and improve safety. Also, a single pressure relief valve is included to release overpressure developed in the battery system. Essentially circuit boards are an integral part of battery cells in the cell-to-pack technology. Integration happens during the manufacturing phase itself as opposed to the assembly phase in the prior art.

In an embodiment of the presented disclosure, as shown in Fig. 6, a double-sided battery system with sandwich structure is envisioned. The collector plates in this embodiment are placed on either side of the battery cell, for example, plate connecting to the cathodes on top of the cells and the plate connecting anodes at the bottom side, in contrast to the previous arrangement where both the collector plates were positioned on the same side of the battery cells. This sandwich structure makes any component that exists in the battery system solely as a structural support

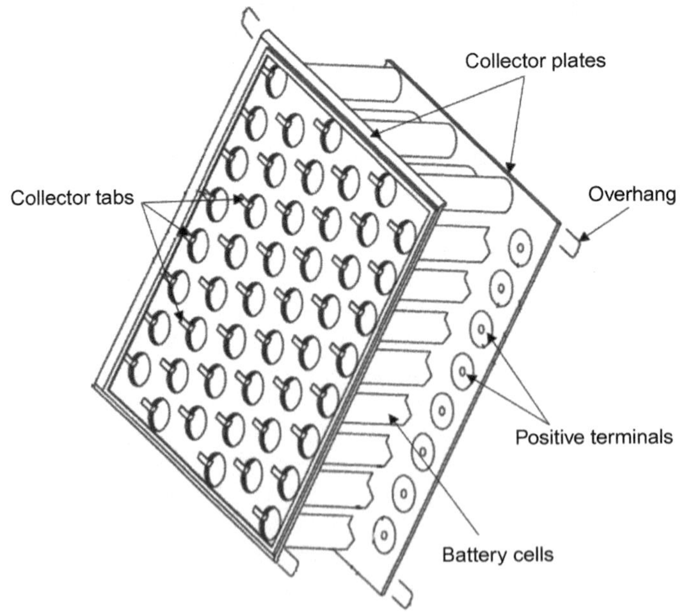

Fig. 6 Illustration showing battery cells sandwiched between the top and bottom collector plates.

member or acts as a fluid conduit redundant. Consequently, container and its associated parts are eliminated from this configuration, thereby providing for a more compact, lightweight packing, and further cost savings [58].

Contemporary Amperex Technology Co., Ltd. (CATL), one of the largest automotive Li-ion battery manufacturer and a pioneer in the cell to pack technology, claims to have increased mass-energy density of Li-ion batteries at the system level from 180 to 200 $Whkg^{-1}$, that is, a 10%–15% increase using this method. Correspondingly, 15%–20% improvements were realized in volumetric density. Overall, the technique resulted in a 40% decrease in the number of parts being used for creating battery packs [59,60]. The cell-to-pack technology offers practical benefits, and other manufacturers are gradually starting to take note of it.

4.4 Summary

Battery packs form the core of electric vehicle technology. This chapter focuses on the two design aspects that are central to engineering reliable battery packs—material selection and manufacturing method.

Batteries with high energy density, high power density, and long cycle life are needed to support the faster introduction and smooth integration of electric vehicles within our society. Choices related to both the materials used in Li-ion batteries and methods used to manufacture the batteries influence their performance. Therefore research is underway to develop novel electrode and electrolyte materials and battery manufacturing methods. Advancements made in materials domain with an aim to develop the next-generation Li-ion batteries are concisely described within this chapter. In addition, advanced manufacturing methods, for example, 3D-printing, the dry electrode method, and cell-to-pack technology, which have the potential to push the development boundaries further are introduced within.

References

[1] D. Andre, S.-J. Kim, P. Lamp, S.F. Lux, F. Maglia, O. Paschos, B. Stiaszny, Future generations of cathode materials: an automotive industry perspective, J. Mater. Chem. A 3 (13) (2015) 6709–6732.

[2] S. Arora, Design of a Modular Battery Pack for Electric Vehicles (Doctoral thesis), Swinburne University of Technology, Melbourne, Australia, 2017.

[3] A. Kulkarni, A. Kapoor, S. Arora, Battery packaging and system design for an electric vehicle, SAE Technical Paper 2015-01-0063, SAE International, Melbourne, Australia, 2015. https://doi.org/10.4271/2015-01-0063.

[4] M. Grimm, L. Lenz, J. Peters, M. Sievert, Demand for off-grid solar electricity–experimental evidence from Rwanda, IZA DP No. 10427, IZA—Institute of Labor Economics, Deutsche Post Foundation, 2017.
[5] S. Arora, A. Kapoor, W. Shen, Application of robust design methodology to battery packs for electric vehicles: identification of critical technical requirements for modular architecture, Batteries 4 (3) (2018) 30.
[6] S. Arora, W. Shen, A. Kapoor, Designing a robust battery pack for electric vehicles using a modified parameter diagram, SAE Technical Paper 2015-01-0041, SAE International, Melbourne, Australia, 2015. https://doi.org/10.4271/2015-01-0041.
[7] S. Arora, A. Kapoor, Mechanical design and packaging of battery packs for electric vehicles, in: G. Pistoia, B. Liaw (Eds.), Behaviour of Lithium-Ion Batteries in Electric Vehicles: Battery Health, Performance, Safety, and Cost, Springer International Publishing, Cham, 2018, pp. 175–200.
[8] S. Arora, W. Shen, A. Kapoor, Critical analysis of open circuit voltage and its effect on estimation of irreversible heat for Li-ion pouch cells, J. Power Sources 350 (2017) 117–126.
[9] Y. Xing, T. Shen, T. Guo, X. Wang, X. Xia, C. Gu, J. Tu, A novel durable double-conductive core-shell structure applying to the synthesis of silicon anode for lithium ion batteries, J. Power Sources 384 (2018) 207–213.
[10] V. Ruiz, A. Pfrang, A. Kriston, N. Omar, P. Van den Bossche, L. Boon-Brett, A review of international abuse testing standards and regulations for lithium ion batteries in electric and hybrid electric vehicles, Renew. Sust. Energ. Rev. 81 (2018) 1427–1452.
[11] A. Shukla, T.P. Kumar, Materials for next-generation lithium batteries, Curr. Sci. 94 (3) (2008) 314–331.
[12] J. Hassoun, F. Bonaccorso, M. Agostini, M. Angelucci, M.G. Betti, R. Cingolani, M. Gemmi, C. Mariani, S. Panero, V. Pellegrini, An advanced lithium-ion battery based on a graphene anode and a lithium iron phosphate cathode, Nano Lett. 14 (8) (2014) 4901–4906.
[13] J.-T. Han, J.B. Goodenough, 3-V full cell performance of anode framework Tinb2o7/Spinel Lini0.5mn1.5o4, Chem. Mater. 23 (15) (2011) 3404–3407.
[14] X. Wen, C. Ma, C. Du, J. Liu, X. Zhang, D. Qu, Z. Tang, Enhanced electrochemical properties of vanadium-doped titanium niobate as a new anode material for lithium-ion batteries, Electrochim. Acta 186 (2015) 58–63.
[15] C. Lin, S. Yu, S. Wu, S. Lin, Z.-Z. Zhu, J. Li, L. Lu, Ru0.01Ti 0.99Nb2O7 as an intercalation-type anode material with a large capacity and high rate performance for lithium-ion batteries, J. Mater. Chem. A 3 (16) (2015) 8627–8635.
[16] H. Song, Y.-T. Kim, A Mo-doped Tinb2O7 anode for lithium-ion batteries with high rate capability due to charge redistribution, Chem. Commun. 51 (48) (2015) 9849–9852.
[17] T.P. Kumar, R. Ramesh, Y. Lin, G.T.-K. Fey, Tin-filled carbon nanotubes as insertion anode materials for lithium-ion batteries, Electrochem. Commun. 6 (6) (2004) 520–525.
[18] J. Chen, Y. Liu, A.I. Minett, C. Lynam, J. Wang, G.G. Wallace, Flexible, aligned carbon nanotube/conducting polymer electrodes for a lithium-ion battery, Chem. Mater. 19 (15) (2007) 3595–3597.
[19] W.-J. Zhang, A review of the electrochemical performance of alloy anodes for lithium-ion batteries, J. Power Sources 196 (1) (2011) 13–24.
[20] H. Wu, Y. Cui, Designing nanostructured Si anodes for high energy lithium ion batteries, Nano Today 7 (5) (2012) 414–429.
[21] P. Poizot, S. Laruelle, S. Grugeon, L. Dupont, J. Tarascon, Nano-sized transition-metal oxides as negative-electrode materials for lithium-ion batteries, Nature 407 (6803) (2000) 496.

[22] V. Etacheri, R. Marom, R. Elazari, G. Salitra, D. Aurbach, Challenges in the development of advanced Li-ion batteries: a review, Energy Environ. Sci. 4 (9) (2011) 3243–3262.
[23] W. Li, A. Dolocan, P. Oh, H. Celio, S. Park, J. Cho, A. Manthiram, Dynamic behaviour of interphases and its implication on high-energy-density cathode materials in lithium-ion batteries, Nat. Commun. 8 (2017) 14589.
[24] C. Liu, Z.G. Neale, G. Cao, Understanding electrochemical potentials of cathode materials in rechargeable batteries, Mater. Today 19 (2) (2016) 109–123.
[25] A. Manthiram, Materials challenges and opportunities of lithium ion batteries, J. Phys. Chem. Lett. 2 (3) (2011) 176–184.
[26] J. Xu, S. Dou, H. Liu, L. Dai, Cathode materials for next generation lithium ion batteries, Nano Energy 2 (4) (2013) 439–442.
[27] M.J. Armstrong, A. Panneerselvam, C. O'Regan, M.A. Morris, J.D. Holmes, Supercritical-fluid synthesis of Fef2 and Cof2 Li-ion conversion materials, J. Mater. Chem. A 1 (36) (2013) 10667–10676.
[28] T. Tsujikawa, K. Yabuta, T. Matsushita, T. Matsushima, K. Hayashi, M. Arakawa, Characteristics of lithium-ion battery with non-flammable electrolyte, J. Power Sources 189 (1) (2009) 429–434.
[29] H. Ota, A. Kominato, W.-J. Chun, E. Yasukawa, S. Kasuya, Effect of cyclic phosphate additive in non-flammable electrolyte, J. Power Sources 119 (2003) 393–398.
[30] H. Nakagawa, Y. Shibata, Y. Fujino, T. Tabuchi, T. Inamasu, T. Murata, Application of nonflammable electrolytes to high performance lithium-ion cells, Electrochemistry 78 (5) (2010) 406–408.
[31] Q. Li, J. Chen, L. Fan, X. Kong, Y. Lu, Progress in electrolytes for rechargeable Li-based batteries and beyond, Green Energy Environ. 1 (1) (2016) 18–42.
[32] M. Ishaq, M. Jabeen, W. Song, L. Xu, Q. Deng, 3d hierarchical Ni2+/Mn2+/Al3+ layered triple hydroxide@ nitrogen-doped graphene wrapped hybrids on nickel foam for supercapacitor applications, J. Electroanal. Chem. 804 (2017) 220–231.
[33] S. Arora, W. Shen, Advances in rechargeable lithium ion batteries and their systems for electric and hybrid electric vehicles, in: T. Maiyalagan, P. Elumalai (Eds.), Rechargeable Lithium-Ion Batteries: Trends and Progress in Electric Vehicles, CRC Press, 2020 (Chapter 3), pp. 99–126.
[34] W. Xing, High energy/power density, safe lithium battery with nonflammable electrolyte, ECS Trans. 85 (13) (2018) 109–114.
[35] G.L. Xu, R. Amine, A. Abouimrane, H. Che, M. Dahbi, Z.F. Ma, I. Saadoune, J. Alami, W.L. Mattis, F. Pan, Challenges in developing electrodes, electrolytes, and diagnostics tools to understand and advance sodium-ion batteries, Adv. Energy Mater. 8 (14) (2018) 1702403.
[36] J. Smekens, R. Gopalakrishnan, N.V.D. Steen, N. Omar, O. Hegazy, A. Hubin, J. Van Mierlo, Influence of electrode density on the performance of Li-ion batteries: experimental and simulation results, Energies 9 (2) (2016) 104.
[37] D.L. Wood, J. Li, C. Daniel, Prospects for reducing the processing cost of lithium ion batteries, J. Power Sources 275 (2015) 234–242.
[38] C. Meyer, H. Bockholt, W. Haselrieder, A. Kwade, Characterization of the calendering process for compaction of electrodes for lithium-ion batteries, J. Mater. Process. Technol. 249 (2017) 172–178.
[39] H. Duong, J. Shin, Y. Yudi, Dry electrode coating technology, in: 48th Power Sources Conference, 2018.
[40] P. Mitchell, L. Zhong, H. Vincent, C. Nanjundiah, Particle Based Electrodes and Methods of Making Same, Google Patents, 2012.
[41] M. Al-Shroofy, Q. Zhang, J. Xu, T. Chen, A.P. Kaur, Y.-T. Cheng, Solvent-free dry powder coating process for low-cost manufacturing of Lini1/3mn1/3co1/3o2 cathodes in lithium-ion batteries, J. Power Sources 352 (2017) 187–193.

[42] C. Morris, Maxwell's Teflon-Fibrilizing Electrode Process Could Save Tesla Big Bucks on Battery Manufacture, 2019, (cited 2020 July 25). Available from: https://chargedevs.com/newswire/maxwells-teflon-fibrilizing-electrode-process-could-save-tesla-big-bucks-on-battery-manufacture/.

[43] M. Vinkhuyzen, Tesla Successful in Buying Maxwell Technologies, 2019, cited 2020 July 25. Available from: https://cleantechnica.com/2019/05/16/tesla-successful-in-buying-maxwell-technology/.

[44] J. Dittmann, N. Willenbacher, Micro structural investigations and mechanical properties of macro porous ceramic materials from capillary suspensions, J. Am. Ceram. Soc. 97 (12) (2014) 3787–3792.

[45] A.H. Slocum, T. Doherty, R. Bazzarella, J.C. Cross III, P. Limthongkul, M. Duduta, J. Disko, A. Yang, T. Wilder, W.C. Carter, Stationary Semi-Solid Battery Module and Method of Manufacture, 2020. Google Patents.

[46] M. Singh, H.M. Haverinen, P. Dhagat, G.E. Jabbour, Inkjet printing—process and its applications, Adv. Mater. 22 (6) (2010) 673–685.

[47] F. Zhang, M. Wei, V.V. Viswanathan, B. Swart, Y. Shao, G. Wu, C. Zhou, 3d printing technologies for electrochemical energy storage, Nano Energy 40 (2017) 418–431.

[48] S. Lawes, Q. Sun, A. Lushington, B. Xiao, Y. Liu, X. Sun, Inkjet-printed silicon as high performance anodes for Li-ion batteries, Nano Energy 36 (2017) 313–321.

[49] J. Hu, Y. Jiang, S. Cui, Y. Duan, T. Liu, H. Guo, L. Lin, Y. Lin, J. Zheng, K. Amine, 3d-printed cathodes of Limn1 − Xfexpo4 nanocrystals achieve both ultrahigh rate and high capacity for advanced lithium-ion battery, Adv. Energy Mater. 6 (18) (2016) 1600856.

[50] C.A. Milroy, S. Jang, T. Fujimori, A. Dodabalapur, A. Manthiram, Inkjet-printed lithium–sulfur microcathodes for all-printed, integrated nanomanufacturing, Small 13 (11) (2017) 1603786.

[51] J. Li, M.C. Leu, R. Panat, J. Park, A hybrid three-dimensionally structured electrode for lithium-ion batteries via 3d printing, Mater. Des. 119 (2017) 417–424.

[52] M. Cheng, Y. Jiang, W. Yao, Y. Yuan, R. Deivanayagam, T. Foroozan, Z. Huang, B. Song, R. Rojaee, T. Shokuhfar, Elevated-temperature 3d printing of hybrid solid-state electrolyte for Li-ion batteries, Adv. Mater. 30 (39) (2018) 1800615.

[53] A. Izumi, M. Sanada, K. Furuichi, K. Teraki, T. Matsuda, K. Hiramatsu, H. Munakata, K. Kanamura, Development of high capacity lithium-ion battery applying three-dimensionally patterned electrode, Electrochim. Acta 79 (2012) 218–222.

[54] C.C. Ho, J.W. Evans, P.K. Wright, Direct write dispenser printing of a zinc microbattery with an ionic liquid gel electrolyte, J. Micromech. Microeng. 20 (10) (2010) 104009.

[55] T.S. Wei, B.Y. Ahn, J. Grotto, J.A. Lewis, 3d printing of customized Li-ion batteries with thick electrodes, Adv. Mater. 30 (16) (2018) 1703027.

[56] K. Sun, T.S. Wei, B.Y. Ahn, J.Y. Seo, S.J. Dillon, J.A. Lewis, 3d printing of interdigitated Li-ion microbattery architectures, Adv. Mater. 25 (33) (2013) 4539–4543.

[57] K. Fu, Y. Wang, C. Yan, Y. Yao, Y. Chen, J. Dai, S. Lacey, Y. Wang, J. Wan, T. Li, Graphene oxide-based electrode inks for 3d-printed lithium-ion batteries, Adv. Mater. 28 (13) (2016) 2587–2594.

[58] K.C. Matthews, Aggregated Battery System, 2019. Google Patents.

[59] PRNewswire, Catl to Drive New Wave of E-Mobility With Next-Generation Electric Vehicle Battery Technology News—Company Reveals Cell-to-Pack Battery Platform at IAA 2019, 2019. (cited 15 July 2020) Available from: https://www.prnewswire.com/news-releases/catl-to-drive-new-wave-of-e-mobility-with-next-generation-electric-vehicle-battery-technology-300915309.html.

[60] Xinhuanet, China's Catl Unveils Cell-to-Pack Battery Platform, 2019, (cited 2020 July 15). Available from: http://www.xinhuanet.com/english/2019-09/13/c_138389934.htm.

CHAPTER 5

EV Battery Pack Engineering—Electrical Design and Mechanical Design

5.1 Introduction

Range anxiety was one of the major issues during the initial development phase of electric vehicles (EVs). It refers to a psychological phenomenon, which makes the rider doubt the energy storage capacity available on-board an EV and question whether it would be enough to complete the journey. Consequently, longer "driving ranges" became a major competing point among the EVs sold commercially. Soon after, it became clear that gains made in the driving range were not linearly proportional to the amount of additional energy storage capacity installed onboard EVs owing to mass compounding effect. It is understood that systems are only as good as their design. Hence rechargeable battery engineering became vital for developing next-generation EVs.

5.2 Battery pack sizing and architecture

Like any other new product development project, the first and the most crucial step in designing a new battery pack is gaining perspective and developing a complete understanding of customers' requirements. These can vary significantly depending on the application. For this reason, Mateusz Figaszewski, Head of E-Mobility Development at Solaris, says, "The bus industry today requires manufacturers to be highly adaptable to technical requirements and the quantities required" [1]. Regardless, all battery packs, at minimum, are expected to have capacities large enough to supply the energy needed to complete a prespecified drive cycle and deliver the power necessary for meeting the peak requirement(s). International standard IEC 61982:2012 defines that the maximum or peak power (P_{peak}) is delivered by a current that causes the open circuit voltage to drop by one-third of its initial value, that is, current that causes

Heavy-duty Electric Vehicles
https://doi.org/10.1016/B978-0-12-818126-3.00004-X

the terminal voltage to become two-third of open circuit voltage instantaneously:

$$P_{\text{peak}} = \frac{2}{3} V_{\text{OCV}} \times I_{\text{peak}} \tag{1}$$

where, V_{OCV} is the open circuit voltage, and I_{peak} is the current that delivers the maximum power [2].

Travel demands can differ with customers, and therefore customizing battery packs as per individual customer's needs is the recommended method for avoiding "oversizing" of battery packs [3]. Uncertainty embedded in end-users "anticipated" behaviour makes the initial battery pack size calculations for a passenger EV challenging [4]. Public transit systems, for example, city buses and waste collection trucks, on the other hand, have a fixed route and schedule.

It is quite usual to have a travel circuit length of approximately 200 km per vehicle per day for city buses. However, the e-bus fleet operated by West Covina's Foothill Transit Group in California runs an average of 305 km per vehicle per day. Timetables of 300 EBUSCO e-buses in Europe require them to drive 325 km per day per vehicle. Similarly, VDL Citea e-buses operate 24 h a day on bus routes around Schiphol, Amsterdam and accumulate travel distance over 100,000 km per year. Dopravný podnik Bratislava (DPB) signs a performance contract with e-bus manufacturers. The contract allows DPB to collect penalties from vehicle manufacturers in case the driving range of any bus their fleet in real-world conditions falls short of the stipulated range of 160 km. For freight transportation, European Union (EU) regulation 561/2006 mandates a compulsory rest period after every 4.5 h of driving. In 4.5 h, the trucks can easily travel 350 km [5]. Therefore e-truck manufacturers are regularly unveiling models with a driving range close to 300 km. Table 1 presents some of the aspects of the commercially available e-trucks.

It is evident from Table 1 that an increasingly wide selection of electric trucks (e-trucks) with a GVW less than 16 tonnes and a driving range in excess of 200 km is reaching the market.

Driving patterns for heavy-duty vehicles can also be regulated to an extent through training—making the battery pack size calculation a bit simple. It is a basic understanding that interrelated subsystems are put together around a Li-ion battery cell to create a functional battery module, as shown in Fig. 1.

A battery module is a block of several battery cells connected in series and in parallel. Series connections are used to meet voltage demands of the

Table 1 Key features of commercial e-trucks available in the market (updated on the basis of information available from [6]).

Manufacturer	Model	Type	Max weight (tonnes)	Battery capacity (kWh)	Range (km)	Energy consumption (kWh/km)	Charging power (AC/DC) (kW)
Mitsubishi	eCanter	Medium duty	7.5	82.8	120	0.69	
BYD	T7	Medium duty	11	175	200	0.88	100/150
Daimler	eM2 106	Medium duty	12	325	370	0.88	260
Volvo	FL Electric	Rigid	16	100–300	100–300	1.0	22/150
Renault	D Z.E.	Rigid	16	200–300	300	1.0	22/150
eMoss	EMS18	Rigid	18	120–240	100–250	1.0	22/44
DAF	LF Electric	Rigid	19	222	220	1.0	300
Mercedez-Benz		Rigid	26	212	200	1.06	
Renault	D Wide Z.E.	Rigid	26	200	200	1.0	22/150
Tesla	Semi	Semitrailer	36		480–800	< 1.25	
BYD	T9	Semitrailer	36	350	200	1.75	100/150
DF	CF Electric	Semitrailer	37	170	100	1.7	300
Daimler	eCascadia	Semitrailer	40	550	400	1.38	260

AC, alternating current; *DC*, direct current.

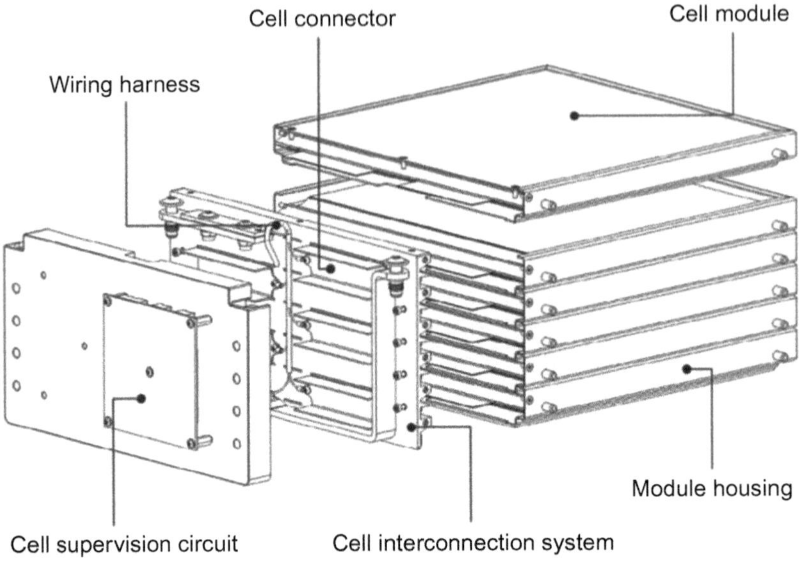

Fig. 1 Battery module based on pouch cells [7] (Reprinted with permission).

system, whereas a parallel circuit amplifies the current handling capability of the network. Accordingly, the number of cells in series in a battery pack or a module can be calculated as:

$$\text{Number of battery cells in series} = \frac{\text{Target pack voltage}}{\text{Nominal cell voltage}} \tag{2}$$

wherein, the target value is based on the electric drivetrain requirements, that is, electric motor specifications. Calculations are rounded off to the nearest even number in order to have a common module design across the battery pack. It should be noted that the recommended practice is to limit module voltages to 60 V. This is because systems delivering more than 60 V are categorized as high-voltage systems and require special tooling plus handling procedure for safe working, adding to the total ownership cost.

The primary function of the battery pack can be described as to hold the battery cells and associated electrical components (battery management system and hardware) and to provide the intended voltage and current. Therefore the pack design encompasses both the electrical architecture and connection strategy and the mechanical design and challenges associated with holding a large number of individual cells in a "pack" arrangement.

Fig. 2 presents a block representation of a typical EV. It shows different electromechanical subsystems, for example, electric motors; electric heating,

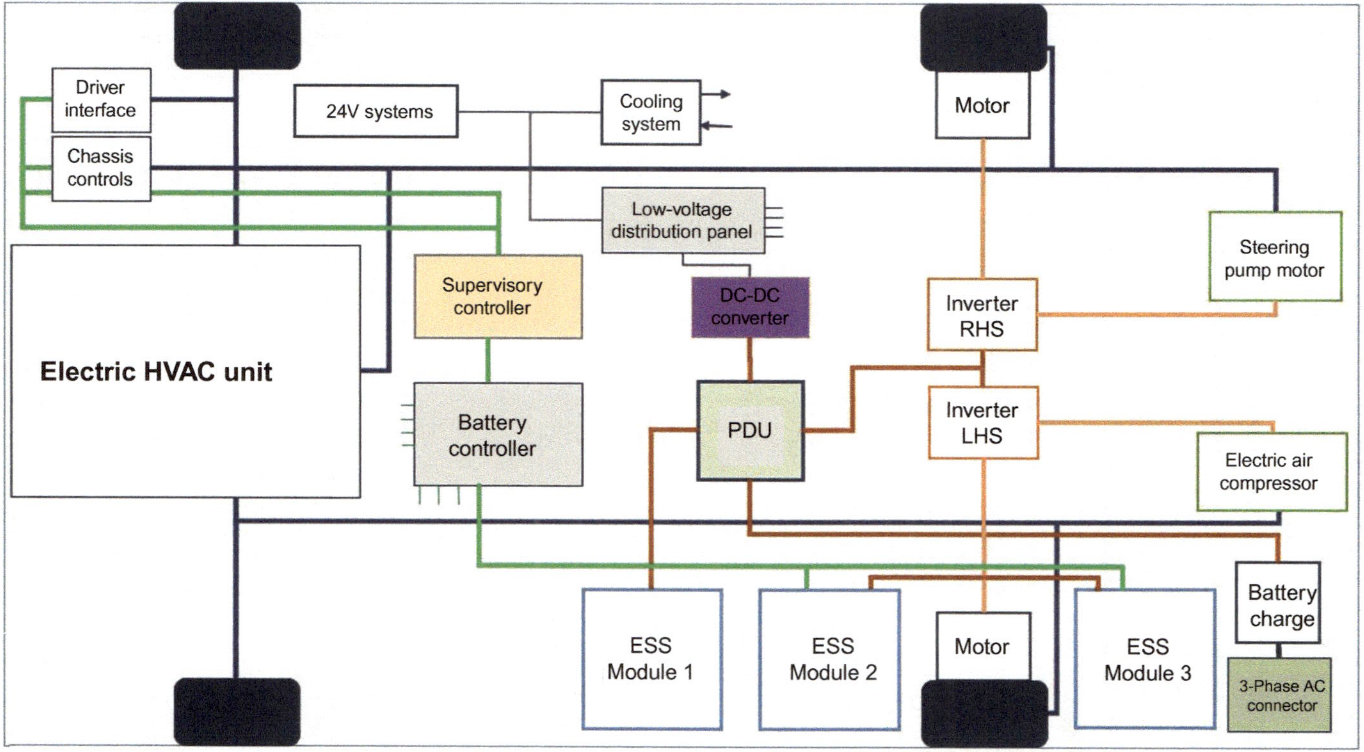

Fig. 2 Schematic of an electric vehicle showing various interfaces. *LHS*, left-hand side; *PDU*, power distribution unit; *RHS*, right-hand side.

ventilation, and cooling (HVAC) unit; electric air compressor; and various auxiliary and control systems that act as load on the on-board energy storage system (EES).

For reliable service, it becomes essential to identify the power requirements of each of the subsystems installed in the EV before defining minimum capacity and power requirements from the battery pack. As an example, an electric bus with a ZF AVE 130 electric axle [8] is considered. In this case, on-board ESS is required to cater to the following subsystems:

- Electric powertrain requires:
  - Peak power of 125 kW per wheel.
  - Power of 87 kW per wheel for 30 min.
  - Continuous power of 60 kW per wheel.
  - Nominal DC voltage of 650 V.
  - Continuous operating current 250 A (AC).
  - Peak current for 30 min 340 A (AC).
- HVAC module: The choice of HVAC is important in the design of an EV because this is the second highest power demand component in EVs. HVAC systems are used solely to maintain operator and passenger comfort independent of the conditions outside the bus. The operational climate for the vehicle is then the biggest driver of the requirements for this subsystem. In tropical climates, one key element of the weather is a high average humidity almost throughout the year, which requires dehumidification to improve comfort levels. It is reasonable to expect that EVs operating in regions with average temperatures of 28°C and above should be capable of providing consistent cooling at some level, whereas cabin heating becomes necessary when the temperatures drop below 20°C. In addition, chillier (subzero) weathers create requirements for preheating of the battery packs themselves. Suddenly, it is not so difficult to appreciate the results of previous research, which confirmed that the power needed for managing the cabin thermal loads under different ambient temperatures reduces the driving range of EVs by 35%–50% [9,10].

  Reliability, cooling performance, and dehumidifying are all high priorities for this subsystem. The performance of this unit then determines the energy consumption and heating and cooling effects on the vehicle; the more efficient the HVAC unit, the less energy consumption for the same comfort level.
- Compressed air supply: Doors, brakes (including parking brakes), and the air suspension systems require compressed air for operation. An

appropriately sized air compressor, driven by an electric motor, is therefore required onboard EVs.

Compressed air systems include supply lines, pressure relief devices, and wet tanks. In an internal combustion engine-driven vehicle, compressors run continuously to maintain a nominal pressure of approximately 7 bars. Excess pressure is bled off to the atmosphere to regulate pressure around the nominal value. In contrast, electrically driven compressors run on-demand because they are typically decoupled from the propulsion motor. As a result, there are no energy losses owing to the bleed-off system. Another major advantage of electrical compressors is a quieter drivetrain and therefore increased comfort for passengers.

In addition, a back-up compressor is connected to the 12 V auxiliary batteries to fill the air tanks and generate operating pressure in the event of a breakdown or when the primary battery pack gets depleted and the bus needs to be towed. To avoid unnecessary operation, the compressor is automatically cut-off when a pressure equivalent to 9 bars is developed.

- 24 V electrical loads: These loads are provided through a DC/DC converter that converts high voltage (HV) energy from the battery pack to 24 V.
- Cooling system: It provides cooling for all HV systems that require liquid cooling.
- Power steering: A standard city bus power steering system is assisted by a hydraulic powered loop that operates a steering gear based on inputs from the steering wheel. The steering gear assists in the movement of the wheels through a "drop-link" to meet the operator's requirements with the minimal physical effort required at the steering wheel. A pump that is mounted to a diesel engine and runs proportionally to engine speed generates the hydraulic power in the system. This operational condition means the steering system must be able to operate fully at all engine speed inputs, which leads to wasted energy at higher engine revolutions per minute (RPM). To adapt this system for a pure EV application, an electrically driven hydraulic pump is required to provide hydraulic pressure for the power-assisted steering.
- Road grade load (%): Driving over high-altitude grades requires more kilowatts per kilometer traveled. Moreover, driving patterns' researches indicate that EVs are driven more frequently in rural and urban topography than on expressways. Consequently, they have to be designed for a more strenuous and demanding environment [11]. Road grade load contributes significantly to the power demand, as discussed in Chapter 3.

Now, Kirchhoff's current rule specifies that at any node or junction, the current coming out is equal to the sum of the currents flowing into it. Accordingly, in a battery module:

$$\text{Number of parallel strings} = \frac{\text{Load requirement}}{\text{Maximum constant current for an individual cell}} \quad (3)$$

5.3 Electrical design

When it comes to electrical network design, EVs present a unique challenge. Operational voltage levels, discussed in Section 5.2, are predetermined based on the powertrain requirements. Higher voltage levels correspond to lower currents. Low current levels equate to cost benefits, mainly from energy distribution system, including cables (smaller cross-sectional areas are needed for low currents), connectors, fuse, etc. On the other hand, low operational voltage leads to savings in battery pack design owing to the need for a reduced number of cell connections and a simpler battery management system for a pack with a smaller number of battery cells. A holistic approach is preferred in the selection of optimal system voltages.

Both AC and DC voltages can be divided into three classes, as shown in Table 2. Voltages lower than 30 V AC or 60 V DC are categorized as Class 1 or low voltage. Electrical isolation of Class 1 circuits from other electric circuits can be ensured without requiring special protection. In this category, traditional 12/24 V level, used for powering convenience features, is the most common in automotive applications. In addition, a new 48-V low voltage level is gradually being introduced to power the electrical components rated at or more than 3 kW, for example, start-stop function, heater, electric air compressor, steering drive, etc.

Last, loads larger than 12 kW are powered through a HV system. Although the specified power demand technically refers to voltage range $60 < U \leq 1500$ V DC or $30 < U \leq 1000$ V AC rms, electric powertrain for heavy-duty EVs is typically designed to operate near 850 V, that is, within voltage class 2. It is governed by the international standard

Table 2 AC and DC voltage classes.

Voltage	Class	AC		DC
		U_{RMS}	U_{Peak}	U_{DC}
Low voltage	1	$\leq$ 30 V	$\leq$ 42 V	$\leq$ 60 V
High voltage	2	$\leq$ 600 V	$\leq$ 849 V	$\leq$ 900 V
High voltage	3	$\leq$ 1000 V	$\leq$ 1414 V	$\leq$ 1500 V

EN50110 and falls under protection class II. Battery packs for passenger cars may operate near 400 V. Unlike 12/24 V level, this protection class poses increased safety risks and can prove fatal in case no additional protective measures are applied. Examples of the safety requirements include sealing to prevent short circuits and arcing, shielding to protect against electromagnetic interference (EMI), and the incorporation of a high voltage interlock loop or HVIL. Therefore in adherence to international standard SAE J1673, all live wires and cables are colour-coded as orange to facilitate easy identification.

The basic hardware set is the same for both the low-voltage system and the HV system. It comprises contactor(s), cell interconnect board, fuses, and busbars. A contactor is a mechanically driven, hermetically sealed switch—opening and closing of which controls the flow of current in an electrical circuit. Enclosing the contacts in a nonconductive environment creates required sealing. Furthermore, the enclosure is filled with a nonconductive gas that minimizes the risk of sparking when the switches open. In EVs, a secondary contactor and a resistor called precharge circuit are connected in parallel to the main contactor. It safeguards the circuitry from damage owing to a sudden inrush of high charge current of motor controller capacitor as the main contactor closes. These contactors and precharge resistors are located in HV power distribution unit (HV-PDU) between the battery pack and the electric motor. There is an interlock loop in the HV connectors. The HVIL includes several contacts in series that create a closed circuit and confirms that all HV connectors are fully connected. If the HVIL line is broken, then supervisory controller stops vehicle operation dues to HV hazard. There is a manual service disconnect (MSD), an integrated contactor typically connected in the middle of the pack, which disconnects the HV from the system. MSD must be disconnected manually for maintenance purposes. BMS controller recognizes the MSD status and in response, opens all the HV contactors. Table 3 provides information on voltage levels of different subsystems found in EVs of different tonnage.

Having adequate electrical isolation between HV circuits and vehicle chassis is essential from a safety perspective. Safe electrical isolation ensures that the electrical current does not pass through the vehicle chassis. It is a crucial requirement to prevent vehicle occupants, first responders, and rescue workers from electrical shock hazards because any object or person can come in physical contact with the chassis during a crash event and close the electrical circuit. In an EV, we can have both - HV DC bus and HV AC bus. They can be electrically insulated

Table 3 Voltage levels for various subsystems of different EV classes (based on information available from [12]).

Components	Voltage (V)			
	Plug-in hybrids	EVs		
	$7.5 < GVW < 12$ tonnes	GVW <7.5 tonnes	$7.5 < GVW < 12$ tonnes	GVW ≥ 18 tonnes
Drive and charging components				
Electric motor (rated voltage)	300	300	400	600
Inverter (DC/AC)	420	420	420	800
Voltage converter (DC/DC)	420–24	420–12	800–24	800–24
Charger (AC/DC)	3 × 400/420	3 × 400/420	400/420	3 × 400/800
Battery	420	420	420	400/800
Subcomponent power				
Compressor	420	420	420	800
Heater	24	12	24	800–24
Electric pumps	24	12	24	24
Steering (electrohydraulic)	Hydraulic	420–12	420–12	800–24

Energy transfer components				
Traditional on-board system	24	12	24	24
Power distributor	420	420	420	800
Cable	420	420	420	800
Connector	420	420	420	800
Isolating elements	420	420	420	800
Relays/contactors	420	420	420	800
Integrated components				
Power semiconductor	650	650	650	1200
Capacitors	450	450	450	880
Resistors	450	450	450	880
Inductors/motor coils	450	450	450	880
Relays/contactors	450	450	450	880
Fuses	450	450	450	880
Current sensors	24	12	24	24
Position sensors	24	12	24	24
Temperature sensors	24	12	24	24

from each other or they can be electrically connected. In the first case, that is, when the two buses are conductively isolated from one another, a minimum isolation resistance of values 100 and 500 Ω/V of the working voltages for the DC bus and the AC bus, respectively, must be provided between the electrical chassis and the HV bus. Otherwise, that is, if the two buses are electrically connected, a minimum isolation resistance of 500 Ω/V of the working voltage is a prerequisite for the safe operation of the system. However, if all AC HV buses are shielded by either mechanically robust enclosures, such as electronic converter case, motor housings that have durability comparable to the service life of the vehicle or by at least two or more layers of solid insulators, or electrical protective barriers, such as wiring harness, then a minimum isolation resistance of value 100 Ω/V of the working voltage is sufficient between the HV bus and the electrical chassis [13]. Electrical isolation can be estimated from the Ohm's law:

$$\text{Electrical isolation} = \frac{1}{\text{Current (A)}} = \frac{\text{Resistance } (\Omega)}{\text{Voltage (V)}} \tag{4}$$

Please note that the United Nations Global Technical Regulation No. 20 stipulates for all applications where a current flow of at least 0.2 A exists between the electrical chassis and any exposed conductive part, the resistance between them shall be smaller than 0.1 Ω. Moreover, the resistance between any two conductive parts of electrical barriers must be less than 0.2 Ω, if they are: a). exposed, b). can be reached simultaneously and c). are less than 2.5 m apart . This requirement can be met by establishing a connection with welding [14].

5.3.1 Circuit-protection strategy

Excessive currents or short circuits can damage the battery cells and pack. A short circuit coordination test is conducted after finishing the electrical design to validate the settings and the fuse/circuit breaker selection. This test confirms that the fuses installed are of correct rating and can interrupt the available current and function selectively. "Selectivity," in electrical systems, signifies that short-time portions (below 0.1 s) and long-time portions (above 0.1 s) of current-time curves (CTCs) would be selective. It means the fuse that is closest to the fault would break/trip first, thereby maximizing the amount of "left in service" portion of the electrical network. In the majority of the cases although, fuses are not selective "on paper" because the CTCs overlap with the fuses' instantaneous overcurrent (IOC).

Coordinated fusing is used to limit the pack damage. The basic strategy is to desensitize or delay the breaker directly upstream of the circuit breaker right next to the fault. The delay stops the upstream breaker from functioning untill the downstream circuit breaker removes the fault. Now pairings available for selective molded case circuit breakers (MCCBs) are limited. Therefore for implementing a "fully selective" architecture, the number of selective circuit breaker layers is usually limited to three [15]. Accordingly,

$$\text{Fusing current (FI): } \text{FI}_{\text{cell}} > \text{FI}_{\text{module}} > \text{FI}_{\text{pack}}$$

Depending on the supplier, battery cells may not come with in-built fuses. In such cases, interconnecting metal near the cell tabs can be constricted to implement the cell-level fusing. One way of doing this includes stamping cell tabs with a pattern of holes. The proper pattern can be determined through experimentation [16] (Fig. 3).

The 2008 National Electric Code Articles 700, 701, and 708 require that the selected pair of overcurrent protective devices (OCPD) must act in a coordinated manner considering the range of available interrupting currents. Historically, insulated case circuit breakers (ICCBs) and MCCBs have demonstrated good "selectivity" even when their CTCs overlap in the instantaneous range.

The short-circuit current (I_{sc}), used for determining the trip current or the fusing current for the protection devices, is calculated as [17]:

$$I_{sc} = \frac{V_{\text{OCV}}}{\left(R_{\text{batt}} + R_{\text{auxiliary}}\right)} \tag{5}$$

where V_{OCV} represents the open-circuit voltage of the battery, R_{batt} is the equivalent internal resistance of the battery, and $R_{\text{auxiliary}}$ is the resistance of the auxiliary system, including cables and connections and protective devices. From the equation, it follows that short-circuit current for the OCPD decreases as the battery ages, and its open-circuit voltage starts to decrease or the combined resistance increases. Furthermore, distinct values must be calculated for the full window of operation, that is, at different SOCs under different operating temperatures because these can significantly alter the short-circuit current. Caution must be taken to ensure that short-circuit current does not become too low; otherwise, it will not be able to cause tripping of the fuse/circuit breaker fast enough and lead to potential fires.

Regarding spatial arrangement, the HV-PDU is the central point for all electrical energy transfer in the EV, connecting batteries, HV components, and chargers to ensure power is distributed where and when it is required. Preferably, it should be positioned as centrally as possible to reduce excessive cable lengths and therefore reducing system energy losses, especially to the motors that require the largest amount of power. The positioning of the remaining components is then based on proximity to functional systems, such as DC-DC converter closest to the 24 V system distribution, or owing to component requirements such as ingress protection.

5.4 Mechanical design

A robust and reliable mechanical design and battery packaging must isolate the performance and safety risks created owing to issues such as vibration transmission, lack of thermal stability, etc. Restricting relative motion at the cell level, modular level, and at the pack level can eliminate some of the potential risks and reduce the severity of some of the others.

5.4.1 Vibration isolation

Over 10% of in-market durability failures reported in EVs is owing to lack of proper vibration isolation of their battery packs and other electrical and electronic subsystems [4]. It is a fact that vibration affects all components of a moving vehicle. Some of these vibrations may be harmless, whereas some may lead to increased deflection and stress levels in the components, while some may cause critical failure of the part. Rather unfortunately, conventional battery pack structures do not isolate individual cells from the transmitted vibrations. Consequently, dynamic mechanical loads build up near intercellular connections and on the busbar leading to loss of electrical connectivity and fatigue failure. A more severe issue develops if the frequency of the transmitted vibration coincides with the natural frequency of the battery pack structure. Resonance creates large amplitude variations causing interlayer delamination in battery cells, which is one of the main failure modes in battery pack [18,19].

Vibrations are mainly of two types. The first category refers to tactile and visual vibrations and includes vibrations in the frequency range of 0–25 Hz. Vibrations with frequencies between 1 and 4 Hz are mainly generated from vehicle handling, whereas road roughness is the primary cause for vibrations in the frequency range of 4–25 Hz. The second category includes vibrations

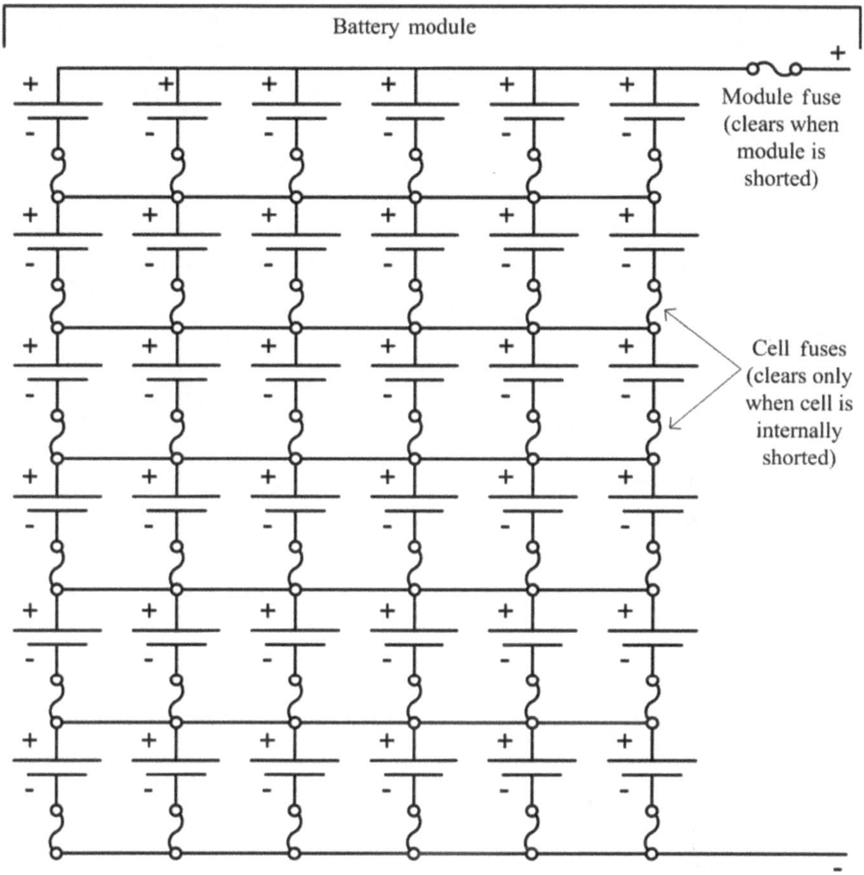

Fig. 3 Fuse layout at modular level.

in the range 25–2000 Hz, which are more commonly referred to as noise [20]. Low-frequency vibrations or the tactile and visual vibrations are regarded as more harmful for battery packs than the noise and thus needs to be isolated. In addition, the top portion of the vehicle body can transmit vibrations with frequencies up to 100 Hz to the battery pack. Therefore it is preferred that natural frequencies of EV battery pack structures exceed 100 Hz.

The natural frequency of any new part that is to be added to the system can be calculated as:

$$f = \frac{1}{2\pi}\sqrt{\frac{k}{m}} \tag{6}$$

where f is the natural frequency, k represents the stiffness of the part, and m indicates its mass.

It follows from the equation that adding more mass or new parts to the system would decrease its natural frequency. It is likely that the new frequency could also belong to the critical range, that is, between 0 and 100 Hz. To reduce the probability, it is essential to make sure that some of the added parts have a higher stiffness than others [21]. Anyhow, two components with equal natural frequencies should not be assembled together in the system. Plus, the use of rubber bushings and damping pads for mounting components is a standard method for dampening amplitude of the transmitted vibrations.

Another working technique for minimizing transmission of vibrations to battery packs involves the application of compressive force through a retainer frame and tensioning bolts. Herein, a rectangular frame structure is used to engage four sides of the battery pack. Tensioning bolts maintain a positive connection between the battery pack and the rectangular frame. In each of the four corners, an L-shaped damping pad is placed, which bears against the frame structure and creates lateral pressure that pushes the battery modules toward one another. In addition, flat compliant pads are arranged between the top and bottom interfaces and in between the adjacent modules. The compliant pads maintain a uniform pressure of 4–18 psi on the cell surface and restrict their movement in *Z*-direction. Besides, they permit small deflection (generally 15%–20%) that accounts for cell expansion during normal (dis)charge operation.

Additionally, weight distribution has a significant effect on the vibration isolation characteristics of a vehicle. Maintaining a low center of gravity offers certain benefits as well. With these two objectives in sight, a battery pack mounting frame was presented in US patent 8561743. The disclosed design makes use of a girder to divide a rectangular mounting frame into two sections—front and rear. The front section is further divided into two parts by a beam welded in the middle of the frame, as seen in Fig. 4. Battery weight is then distributed equally across it by switching the cell orientation from vertical (in the front section) to transverse at the rear [22].

As a rule of thumb, heavier items must be placed at a lower level than the lighter items to keep the center of gravity relatively low. Because battery modules weigh more than typical travel luggage, it is advisable to install battery packs below the luggage compartment. However, owing to the low floor design of modern buses, the majority of the available packaging space is at the rear of the bus. Placing excessive weight on the rear axle limits the passenger-carrying capacity of the e-bus. Therefore priority is given to placing batteries forward of the rear axle to help transfer some of that weight forward, meaning tighter packaging constraints. *Owing to these constraints, BYD places a portion of the on-board battery pack in the middle of the bus and the rest on the top in its e-bus models.*

Last, it is recommended to use intercellular connectors with an in-built positive locking mechanism to minimize the effects of transmitted vibrations on the durability of EV battery packs. It becomes critical for EVs operating in the countryside. Interlocking connectors make sure that the cell connections do not separate over time, and electrical continuity of the battery pack is maintained through its service life.

Table 4 presents applicable standards for testing vibration isolation characteristics of EV battery packs.

Generally, HDVs experience more severe vibration modes than passenger cars. Also, the recorded vibration modes may differ significantly between various HDVs based on their end applications. Hence application-specific vibration profiles are often preferable for compliance testing of heavy-duty EVs. For the tests of HV battery packs to be successful, there must be no evidence of rupture, electrolyte leakage, venting, fire, or explosion. Pieces of evidence (particularly for electrolyte leakage and venting) are collected through visual inspection without detaching the battery pack or any part of it from the vehicle. More importantly, the isolation resistance measurement test performed after the vibration characteristic test must return a value greater than $100\,\Omega/V$.

In addition to vibration isolation, a reliable battery packaging design should minimize thermal and mechanical interactions between different units of the battery pack and address issues relating to thermal stability and impact resistance at the cell level and module level to reduce the probability of battery pack failure.

5.4.2 Thermal stability

An uncontrolled chain of reactions (and ineffective heat dissipation) can push battery cells into a state of "thermal runaway." It is a condition where the cell voltage decreases instantaneously or the cell temperature increases beyond the maximum limit predefined by the cell manufacturer. In either scenario, the self-heating rate of the battery is greater than 1°C/s [23]. It can be initiated during everyday use owing to inherently present manufacturing defects or through unintended abuse or an internal short circuit. It causes a build-up of high pressure inside the cells and makes them emit jets of effluent materials and flammable gases. As a matter of fact, a battery can emit hot fumes, gases, and effluents without

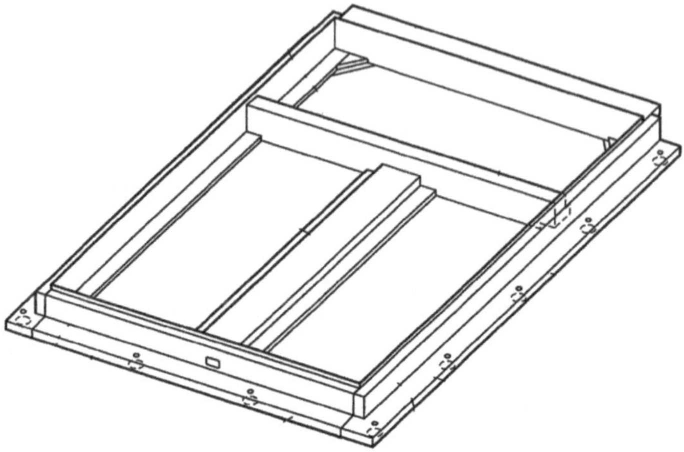

Fig. 4 Perspective view of battery mounting frame [23].

experiencing thermal runaway. High-temperature emissions create safety risks for vehicle passengers, first responders, and nearby property. In extreme events, batteries catch fire and explode spontaneously [24,25]. Fig. 5 shows three possible means of increasing the thermal stability of battery packaging design. They are as follows:

1. Controlling point of egress for hot fumes and effluents.
2. Inclusion of thermal barriers at the cell level and module level.
3. Addition of thermal management system.

It is essential to include at least one pressure release valve or a preidentified failure point, designed to go off at specified pressure, to minimize the safety risks owing to an unknown failure location. Controlling the release direction of hot gases and fumes away from the apparent approach path or the zone of operation enables to limit the safety risks for people and the extent of damage caused to the vehicle/property. Point of egress or the release direction of hot fumes can be controlled by integrating several exhaust nozzles in battery packaging design. Installing thermal barriers strategically at appropriate locations inside the battery packaging can prevent thermal runaway propagation to a larger section of the battery pack. Exhaust nozzle assembly remain closed during regular vehicle operation, and a seal prevents moisture or any road debris from entering into the vehicle. A pressure equalization valve with cracking pressure in the range of 0.5–1.0psi is incorporated into the exhaust nozzle to provide a means for handling pressure to provide means for handling pressure

Table 4 Standards applicable to cell level vibration tests.

Standard	Application	Type of vibration	Frequency (Hz)	Duration	Peak load
UN 38.3 T3	Transport of Li-ion cells	Sine with logarithmic sweep	7–200	3 h	8 g
UN 38.3 T4	Transport of Li-ion cells	Half sine shock	–	18.6 ms	150 g
IEC 62660-2 6.1.1	Electric road vehicles	Random vibrations	10–1000	8 h	27.8 g
IEC 62260-2 6.1.2	Electric road vehicles	Half sine shock	–	20.6 ms	50 g
UL 1642 15	Lithium cells and batteries	Shock	–	18 times	125–75 g
UL 1642 16	Lithium cells and batteries	Harmonic vibrations	10–50	90–100 min	0.8 mm amplitude

differentials owing to non-thermal events, for example, altitude variations. This cracking pressure is much less than the pressure encountered during thermal runaway. Furthermore, battery packaging is designed using hollow structural members whose purpose is to guide the hot fluids and effluents, vented by the battery during thermal runaway events, away from where people and property may get affected by them. Thermal stability of the battery packaging can be further increased by using perforated compartments [26].

Rigidity and integrity of battery packaging get compromised during thermal runaway because mounting brackets, holding the battery cells/modules in their original position, start to melt away owing to high temperatures in the affected region. Intercellular and intermodular spacing is decreased because the affected battery starts moving from its place. This causes a significant reduction in batteries' resistance to thermal runaway propagation. In stacked type configurations, gravitational forces can further expedite the movement of the upper layer once the mounting bracket is deformed and becomes loose. The cells may likely come to rest against the neighbouring cells causing internal short circuit. It is therefore critical to install thermal barriers at multiple levels in order to contain the damage. It can be achieved in several ways as shown in Fig. 5.

First, compartments are created to locate each battery module separately inside the battery pack using crossmembers. A central battery pack member is used to divide each compartment further into left and right sections. Besides allowing easy positioning of each module and holding them in their place, the central member also introduces an air gap between the battery modules and the surface of the battery packaging unit. The air gap inhibits conductive heat transfer from one module to another. Second, friction-fit or bonded, rigid cell spacers can be used to keep the cells in their prespecified location. The type of spacers needed depends on the shape and size of the battery cells employed in the pack. Two small spacers, an upper spacer and a lower spacer, are generally preferred over a one long spacer to save mass. Height of the cell spacer is usually 1%–5% of the battery cell's height. During its lifetime, a battery cell expands 3%–5% of its initial thickness. Besides providing cell-holding functionality, cell spacers create a binding pressure that counters internal spring forces and prevents it from expanding beyond the usual range as it ages [18].

BYD prefers to maintain a 0–0.5 mm intercellular spacing in their battery packs. Furthermore, a foamed material is used as cell spacer or for filling up the intercellular gaps. Third, a thermal management system can be included in the battery

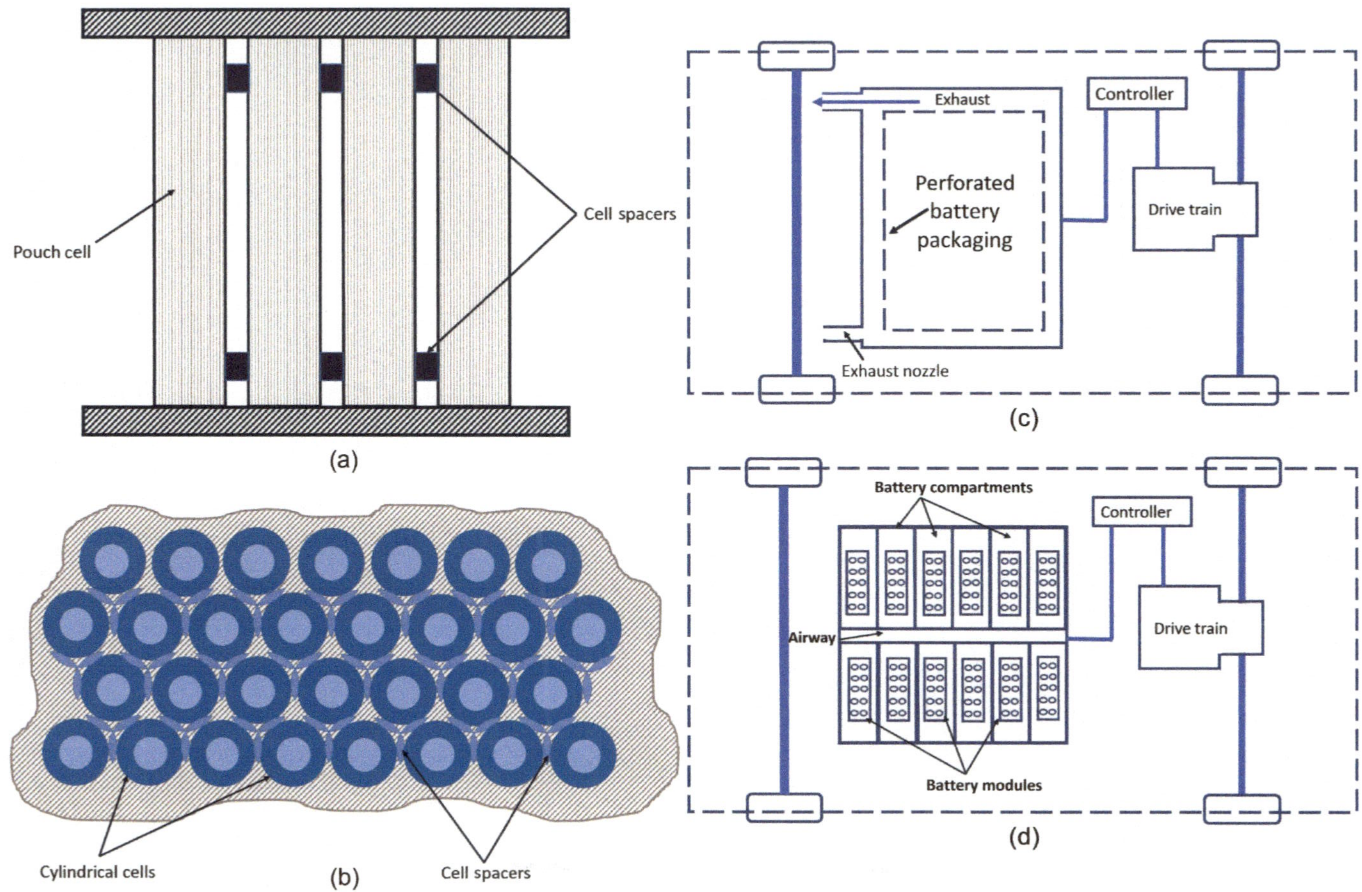

Fig. 5 Thermal barriers in a battery pack (a) cell spacer between pouch cells, (b) section view showing cell spacers between cylindrical cells making the module, (c) exhaust nozzles for controlling venting location, and (d) compartmentalization of battery pack.

packs to maintain the battery cells in a predesignated temperature range [27,28]. Battery thermal management systems will be discussed in detail in Chapter 8.

5.4.2.1 Thermal stability test

If unchecked, thermal runaway can propagate to neighboring cells and modules. Affected cells vent hazardous gases and flammable electrolyte, thereby exposing the vehicle and its occupants to high safety risks. United Nations Global Technical Regulation No. 20 describes a test procedure to assess the propensity of the battery pack for thermal runaway propagation and to ensure the safety of EV occupants. The procedure recommends three methods to initiate thermal runaway in the test device:

1. Nail penetration: Cell is inspected to identify the location and direction of nail penetration that can cause thermal runaway in a cell. A steel nail with a diameter of 3 mm or more and circular cone tip (angle of 20–60 degrees) is inserted in the cell at the identified location at a speed of 0.1–10 mm/s.
2. Overcharge: A constant current, C-rates between 0.3C and 1C, is used to overcharge the cell. It is charged until SOC of 200% is attained or till the cell experiences thermal runaway event, whichever is sooner.
3. Heating: To use this method, modifications need to be made to the battery module by either attaching a film heater to the cell surface or by replacing one of the component cells with a block heater. The heater must be covered with metal, insulator, or ceramic, and the heating area contacting the cell should be equal to the surface area of the cell. If dimensions of the block heater are smaller than that of the component cell, then it is placed in the module that is making surface contact with the test cell. The heater is set ON at the maximum power mode. Heating is stopped if either of the conditions is reached:
   - cell goes into thermal runaway
   - measured cell temperature exceeds 300°C
   - heating is ON for 30 min

The manufacturer can choose any of the three methods for test purposes. The tests can be conducted at vehicle level or at pack level. The manufacturer can also choose to perform the tests at the subsystem level, that is, test battery modules, including cells and associated electrical connections. However, they must provide conclusive evidence that confirms the module performance is representative of the pack performance. Moreover, the tests must be conducted at a temperature of

$25 \pm 2°C$ in an indoor facility to minimize the effect of wind on the test results, and a BMS, if not integrated into cell enclosure, must be configured to send warning signals upon initiation of a thermal event.

If the chosen initiation method fails to stimulate the exothermic reactions required to push the battery into the state of thermal runaway, then the tested device is classified to meet the runaway propagation requirement for the selected method. To ensure complete safety, secondary tests are conducted with the remaining two methods. In case the thermal runaway occurs, time since activation of the thermal runaway warning signal is measured. Visual inspections are made without disassembling the tested device. At pack level, if no external fire appears or any explosion occurs within 5 min of receiving the warning signal, then the pack is considered to meet the safety requirements. At vehicle level, if no smoke enters the passenger cabin, no external fire is witnessed and no explosion is heard within 5 min of sounding the warning alarm for a thermal event, then the EV is classified as safe for its passengers [14].

5.4.3 Battery swapping stations

In BYD buses, battery modules are arranged in the middle section of the chassis. Some modules are placed on the top of the bus as well. In case a faulty cell is detected, BYD engineers prefer to replace the whole module instead of the faulty cell. Maintenance sheet for their K9 model indicates that 8.0 man hours were needed for dismantling modules from the forward interior portion of the battery pack. In contrast, it took 29.0 man hours and 35.0 man hours, respectively, for removing rear battery trays and the rear battery pack [29]. The necessity of frequent battery pack replacement or swapping operations for replenishing energy storage capacities is decreasing with increasing battery densities and a growing network of fast/superchargers. Nevertheless, it makes sense to have quick battery exchange stations readily available to reduce the maintenance downtime.

A system for swapping battery packs in EVs was disclosed by Tesla in US patent number 10513247B2. This system includes a frame with a base and a rack for holding the battery pack and a lift for raising or lowering the frame. Two sets of four air bearings each are also mounted on the frame. The first set is positioned on top of the frame and allows relative movement between the frame and the lift, whereas the second set is positioned on the bottom side of the frame and permits it to move relative to the battery pack. The rack is also moveable relative to the base of the frame.

For swapping to happen, a vehicle drives over a ramp, supported by pillars, and is parked as close to the battery pack lift system as possible. Vehicle guides help in proper positioning, which is critical to the procedure. The frame is then aligned to the battery pack in the vehicle with the help of the air bearings. Extension of an alignment pin, fixed to the frame, further ensures proper alignment between the vehicle and the swapping system. "Proper alignment" refers to a position where the rack touches the underbody of the battery pack in the vehicle and each of the torque-control devices or nut runners, mounted on the rack, is lined up perfectly with individual fasteners used for securing the battery pack to the vehicle body.

The torque-control devices are typically configured according to fastener pattern(s), that is, connector or slot designs specific to particular vehicle types. Subsequently, suitable torque is applied by using nut runners to unfasten the battery pack. The rack is moved at least vertically for assisting in the unfastening process. Point to note is that orienting all the fasteners in the same direction can make the fastening and unfastening process for the battery packs more streamlined. Last, the unfastened pack is rolled off over the rack and lowered/removed from the vehicle's body [30].

Parking a heavy-duty EV next to the battery swap station accurately might prove tricky. It is estimated that approximately 400 m^2 are required for setting up battery pack replacement stations for buses. Chinese Patent CN102904311 presents a battery replacing station for electric vehicles wherein an electric forklift is used to lock the vehicle and for removing the battery from its body. The vehicle thus need not be accurately positioned, which significantly reduces the difficulty in parking heavy-duty EVs in tight spaces (Fig. 6).

In this design, the forklift acts like a front-mounted gantry. Battery extraction mechanism is placed on the front end of the gantry along with a positioning sensor, controlled through an onboard programmable logic controller. Sensor feedback allows the gantry to locate and lock the target using top, down, left, and right movements followed by forward motion. Thereafter, the battery pack is extracted using forklift's built-in DC motor and an electromagnet.

The forklift then places the removed battery pack on the input conveyor belt that transports it to a transfer machine and then to a stacker wherein the battery is RFID-enabled and positioned on the charging platform. On the other side, a fully charged pack is picked up from the stacker and delivered to the battery output port by the returning conveyor. Electric forklift installs the fresh battery pack back in the vehicle thus completing the process [31].

Battery swapping stations are capital-intensive, permanent structures. More importantly, they are vehicle $\Longrightarrow$ battery $\Longrightarrow$ connector-specific. It means they lack adaptability, their usage is difficult to optimize, which makes their expansion a challenge. Notwithstanding, they allow separation of battery cost from the vehicle cost, thereby lowering the acquisition cost of EVs significantly. This is one of the strategies being applied to facilitate the electrification of HDVs in developing countries. US patent number 5927938 disclosed the conceptual design of a modular battery swapping station. The modular approach enabled the construction of battery exchange stations with a low initial investment. Furthermore, modular expansion plans allow the swapping station to achieve maximum productivity. In this design, as an EV enters the servicing area of a battery swapping station, externally powered drive sprockets get coupled to the notches on the bottom surface of the battery pack through slidably engaging electrical connectors. Sprockets engage the pack and extract it from the seat in the battery compartment of the vehicle and shift it onto a conveyor belt with the assistance of a hydraulic ram. Retention structures, such as compartment hatches or vertically extending ridges, either permanently fixed or movable between "locked" and "unlocked" positions are used for locking the battery pack in the seat and minimize the probability of lateral movement during operation. Furthermore, a shoulder on the installation side can be provided to restrict any unwanted movement in the reverse direction. Continuous conveyor loop, between the transfer station and the delivery end, is oriented in a vertical direction to save construction space. Battery chargers are placed along the longitudinal run of the conveyor belt at respective battery resting points for recharging the received battery packs. The conveyor loop is designed by mounting a frame structure over a number of vertical support posts. The support posts extend vertically beyond one frame structure to create an overhang for enabling placement of a series of vertically spaced expansion modules, added as needed to accommodate higher demand. All the expansion modules are identical structures, that is, include a conveyor loop with strategically arranged battery recharging points and operate in exactly the same way as another. Battery elevators are arranged on opposite ends of the loop to facilitate the transfer of battery packs between different conveyor loops [32].

Although a plurality of spaced notches or recesses on the (under)side of the battery housing is preferred for engaging the drive mechanism of the

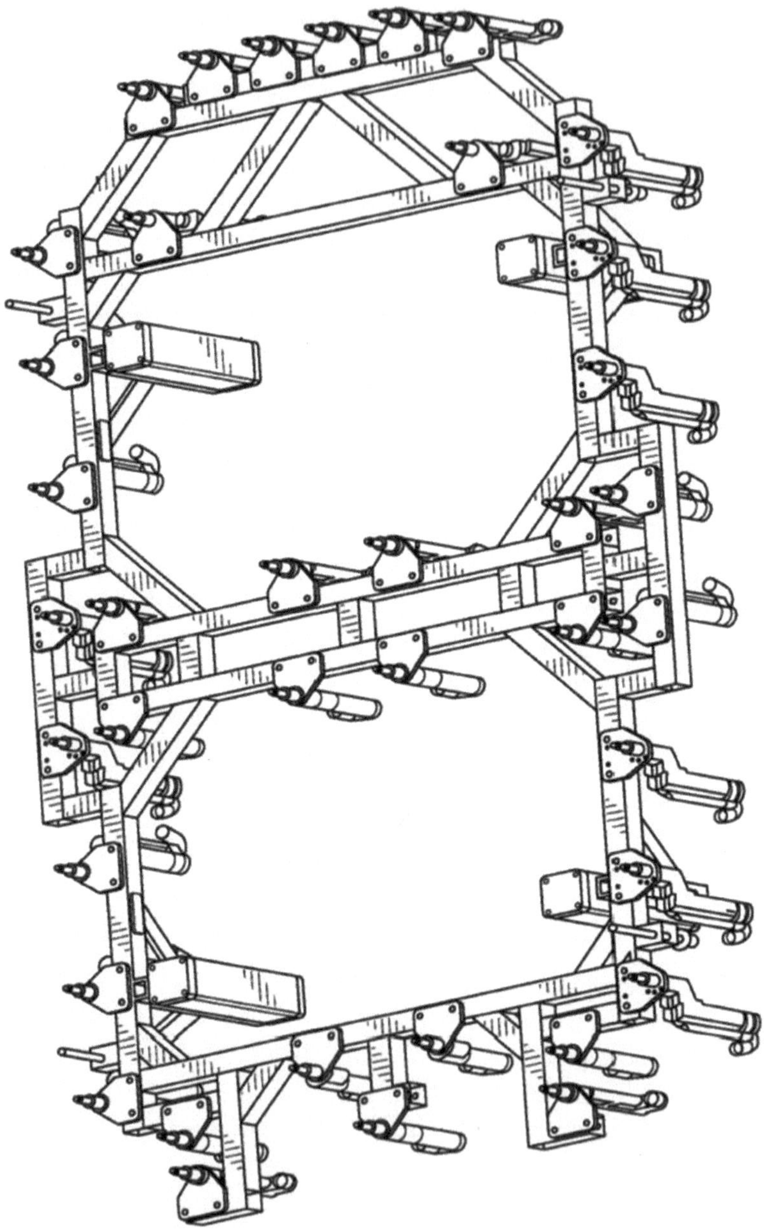

Fig. 6 Battery swapping mechanism with replaceable nut runners.

battery swapping station, other options such as rings, projections, hooks, etc. are equally viable. Choice of engagement structure should be based on the load to be transferred or the weight of the battery pack(s); static friction created by the structural barriers midway, or that should be overcome while replacing the battery pack; and according to the direction of displacement, that is, if exchange process requires moving the battery in the horizontal plane or lifting it in the vertical direction.

Battery contacts come in a different form and allow to reversibly place the battery pack in electrical contact with both the vehicle and the charging station. One type of electrical contacts includes spring-loaded, retractable members. They respond to a physical driving force by retracting into the battery housing. Another type includes conductive contact surfaces that can be placed either above or below the adjoining surface of the battery pack. Conductive cables, clips, pins, and plugs are some other varieties of electrical contacts that can be used.

The design disclosed in the US patent number 8164300 utilizes an exchange platform fitted with a replaceable gripper for handling swapping process of batteries of different types and sizes. The gripper is configured with all the hardware necessary for grasping the battery safely and then exchanging it. A rack, housing grippers of different sizes, is placed adjacent to the exchange platform. Upon receiving information/instructions about the vehicle and the battery type approaching the service area, an automated robotic mechanism retrieves the appropriate gripper from the rack and attaches it to the battery exchange platform. Furthermore, an indexing system equipped with an image processing system is used for locating the battery pack. It measures alignment of the battery compartment of the vehicle with respect to the exchange platform and sends feedback so that the gripper is positioned correctly for replacing the battery. The final alignment is realized through locating pins that align themselves into pilot holes on the (under) side of the battery housing [33].

Effect of EVs on the city's power distribution grid is still highly debated and a matter of concern to many policymakers and fleet managers. Battery swapping stations are expected to keep several battery packs ready for installation in stock. These packs (and more) are charged simultaneously during the storage period at swapping station. Large buffer capacities of the battery swapping station, therefore, allows the station managers to use lower charging currents for the packs or defer the charging event to off-peak periods altogether. Fast charging accelerates battery aging and degradation behaviour. Battery swapping stations, in a way, prolong battery lifetime by enabling controlled charging with low currents and reducing the instances when fast charging becomes unavoidable. Other advantages of battery swapping stations can be summarized as:

- Removing major impediment to large-scale EV adoption.
- Decreasing the total cost of ownership for EV buyers: reduced sales price tag for EV + no compulsion to upgrade household electrical circuit for installing costly high-power EV charger at home.
- Reduce EV inactivity period by eliminating charging-related wait-times.
- Promote the mental well-being of EV riders by providing a practical tool to counter the issue of range anxiety.
- Absorption and flattening of electric grid peak load and minimize network congestion.
- Act as an energy reservoir during emergency grid shutdown.

There are many practical benefits of constructing battery swapping stations. Companies working in this space should, therefore, be encouraged. A couple of points worth noting are: (1) station owners must be mindful of the initial stock and inventory needed to be purchased to make this a sustainable business; and (2) support from OEMs in standardizing the battery packs used by different vehicles would potentially reduce the inventory size.

5.5 Summary

High reliability requirements from EV operation demand utmost attention to engineering of their battery packs. Battery pack engineering involves determining battery pack sizing suitable for targeted application, creating a robust electrical connection network and designing appropriate battery packaging and support systems. Proper design ensures delivery of anticipated performance levels and safe operation under a range of operating conditions. Various design aspects related to the engineering of rechargeable battery packs for heavy-duty EVs are discussed in this chapter.

Elements of HV systems and corresponding electrical isolation needed to contain electric shock hazards are explained. Possible means of limiting transmission of vibration to battery packs and increasing their thermal stability are presented. In addition, the standard procedure for testing the propensity of battery packaging designs to thermal runaway propagation is summarized. In the end, patented mechanisms developed to eliminate lengthy charging periods and to reduce maintenance downtime by swapping depleted (or faulty) battery packs with healthy and charged ones are described.

References

[1] A. Fedorska, What's Driving the Electric Bus Boom in Poland? 2019, Bus Operators Demand Individually Adapted Vehicles (cited 2020 June 28). Available from: https://www.electrive.com/2019/11/13/whats-driving-the-electric-bus-boom-in-poland/.

[2] IEC, Secondary Batteries (Except Lithium) for the Propulsion of Electric Road Vehicles—Performance and Endurance Tests, International Electrotechnical Commission, 2012, p. 79. 61982.

[3] IEA, Global EV Outlook 2019, IEA, Paris, France, 2019. https://www.iea.org/reports/global-ev-outlook-2019.

[4] S. Arora, A. Kapoor, W. Shen, Application of robust design methodology to battery packs for electric vehicles: identification of critical technical requirements for modular architecture, Batteries 4 (3) (2018) 30.

[5] European Parliament, Regulation (EC) No 561/2006 of the European Parliament and of the Council of 15 March 2006 on the Harmonisation of Certain Social Legislation Relating to Road Transport and Amending Council Regulations (EEC) No 3821/85 and (EC) No 2135/98 and Repealing Council Regulation (EEC) No 3820/85, Off. J. Eur. Parliament (2006). L 102/1.

[6] H. Liimatainen, O. van Vliet, D. Aplyn, The potential of electric trucks—an international commodity-level analysis, Appl. Energy 236 (2019) 804–814.

[7] C. Linse, R. Kuhn, Design of high-voltage battery packs for electric vehicles, in: Advances in Battery Technologies for Electric Vehicles, Elsevier, 2015, pp. 245–263.

[8] ZF, Axle & Transmission Systems for Buses & Coaches. Product Overview, 2019. (cited 15 October 2020) Available from: https://www.zf.com/products/en/buses/products_40128.html.

[9] K.R. Kambly, T.H. Bradley, Estimating the HVAC energy consumption of plug-in electric vehicles, J. Power Sources 259 (2014) 117–124.

[10] A. Kulkarni, A. Kapoor, S. Arora, Battery packaging and system design for an electric vehicle, SAE Technical Paper 2015-01-0063, SAE International, Melbourne, Australia, 2015, https://doi.org/10.4271/2015-01-0063.

[11] J.M. Hooper, J. Marco, Characterising the in-vehicle vibration inputs to the high voltage battery of an electric vehicle, J. Power Sources 245 (2014) 510–519.

[12] H.M. Fischer, L. Dorn, Voltage Classes for Electric Mobility, ZVEI - German Electrical and Electronic Manufacturers' Association,, 2013.

[13] UNECE, Regulation No. 100 of the Economic Commission for Europe of the United Nations (UNECE)—uniform provisions concerning the approval of vehicles with regard to specific requirements for the electric power train, Off. J. Eur. Union (2015). L 87/1.

[14] United Nations, Global Technical Regulation No. 20, Electric Vehicle Safety (EVS) (ECE/TRANS/180/Add.20), United Nations, 2018. GE.18-07039(E).

[15] GE-IndustrialSolutions, Guide to Low Voltage Systemdesign and Selectivity, GE Publications Library, 2014. www.geelectrical.com. DET-654C.

[16] A123 Energy Solutions, Battery Pack Design, Validation, and Assembly Guideusing A123 Systems Amp20m1hd-a Nanophosphate®Cells, A123 Systems, LLC, 2014.

[17] Hu, P., Design and specification for safe and reliable battery systems for large ups. Schneider Electric White Papers. White Paper No. 207, 2020. https://www.se.com/hk/en/download/document/SPD_VAVR-9JWR7J_EN/.
[18] S. Arora, W. Shen, A. Kapoor, Review of mechanical design and strategic placement technique of a robust battery pack for electric vehicles, Renew. Sust. Energ. Rev. 60 (2016) 1319–1331.
[19] S. Arora, W. Shen, A. Kapoor, Designing a robust battery pack for electric vehicles using a modified parameter diagram, SAE Technical Paper 2015-01-0041, SAE International, Melbourne, Australia, 2015, https://doi.org/10.4271/2015-01-0041.
[20] T.D. Gillespie, Fundamentals of Vehicle Dynamics, vol. 400, Society of Automotive Engineers, Warrendale, PA, 1992.
[21] F. Jönsson, J. Kindahl, Packaging Concepts of an Energy Storage System for a Fully Electric Heavy Duty Truck (Master's thesis), Chalmers University of Technology, Sweden, 2018. https://hdl.handle.net/20.500.12380/255262.
[22] M. Iwasa, S. Ogata, H. Kadota, T. Hashimura, N. Mori, Vehicle Battery Mounting Structure, Google Patents, 2013.
[23] S. Arora, Design of a Modular Battery Pack for Electric Vehicles (Doctoral Thesis), Swinburne University of Technology, Melbourne, Australia, 2017.
[24] S. Arora, Selection of thermal management system for modular battery packs of electric vehicles: a review of existing and emerging technologies, J. Power Sources 400 (2018) 621–640.
[25] S. Arora, A. Kapoor, Experimental study of heat generation rate during discharge of Lifepo4 pouch cells of different nominal capacities and thickness, Batteries 5 (4) (2019) 70.
[26] S. Arora, A. Kapoor, Mechanical design and packaging of battery packs for electric vehicles, in: G. Pistoia, B. Liaw (Eds.), Behaviour of Lithium-Ion Batteries in Electric Vehicles: Battery Health, Performance, Safety, and Cost, Springer International Publishing, Cham, 2018, pp. 175–200.
[27] S. Arora, A. Kapoor, W. Shen, A novel thermal management system for improving discharge/charge performance of Li-ion battery packs under abuse, J. Power Sources 378 (2018) 759–775.
[28] S. Arora, W. Shen, A. Kapoor, Neural network based computational model for estimation of heat generation in Lifepo4 pouch cells of different nominal capacities, Comput. Chem. Eng. 101 (2017) 81–94.
[29] Bus Testing and Research Center, Byd Electric Bus—Federal Transit Bus Test Performed for the Federal Transit Administration U.S. Dot in Accordance with Cfr 49, vol. 7, 2014. Part 665.
[30] A. Clarke, M.L. Brown, E.O. Gaffoglio, I.S.M. Poznanovich, Battery Swapping System and Techniques, Google Patents, 2019.
[31] H.Q. Li Xiangzhen, O. Qinghai, Z. Lingkang, Z. Yan, J. Lihua, Electric Vehicle Battery Sensing Device, Electric Vehicle Battery and Charge and Discharge Method, European Patent Office, China, 2013.
[32] J.G. Hammerslag, Battery Charging and Transfer System for Electrically Powered Vehicles, Google Patents, 1999.
[33] S. Agassi, Y. Heichal, Battery Exchange Station, Google Patents, 2012.

CHAPTER 6

Charging Technologies and Standards Applicable to Heavy-duty Electric Vehicles

6.1 Introduction

As the name suggests, electric vehicles (EVs) operate with electrical power that could be generated internally or retrieved from an energy storage device. Fuel cell-powered EVs fall into the "internally generated" category, while battery-powered EVs fall into the energy storage category. Even though there are growing interests in fuel cell-powered EVs, battery-powered EVs are the most common and technically matured types of EVs available in the market. In battery-powered EVs, storing energy in batteries and retrieving them are the most important aspects because they are the fundamental processes that give the energy to operate the vehicle. These processes are known as charging and discharging of EV batteries. Of the two processes, charging is where the user interaction is required and often the crucial factor that determines the convenience and confidence in planning a journey.

Commonly used battery types in EVs belong to the Li-ion family. However, the exact chemistry used in a particular EV battery type may vary depending on the materials used in the anode, cathode, and electrolyte. Irrespective of the chemistry, charge/discharge mechanism is fundamentally based on the movement of Lithium ions through the electrolyte and the separator. The movements of Lithium ions during charging and discharging processes are graphically illustrated in Fig. 1a and b, respectively.

During the charging process, the charger provides a sufficiently large negative voltage at the anode. In other words, the charger supplies high-energy electrons to the anode (note: as marked in Fig. 1, direction of current is opposite to the direction of electron flow). These high-energy electrons cause the Lithium atoms in the cathode area to become Lithium ions by releasing an electron. The corresponding reaction at the cathode is

Heavy-duty Electric Vehicles
https://doi.org/10.1016/B978-0-12-818126-3.00008-7

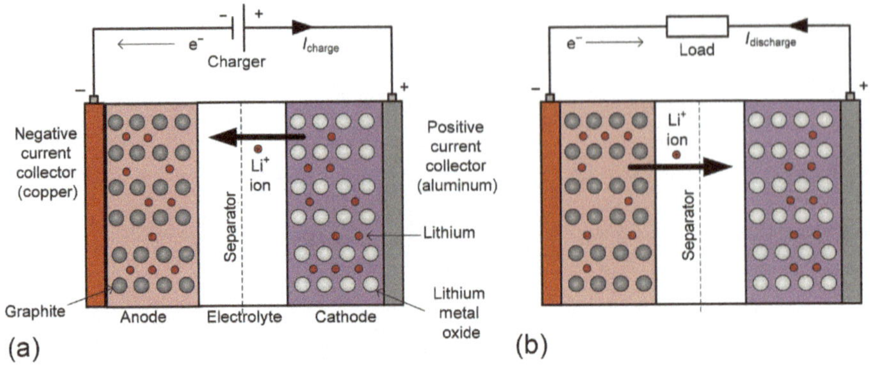

Fig. 1 (a) Charging process and (b) discharging process of Li-ion batteries.

given in Eq. (1), where LiMO is a Lithium transition metal oxide, for example, LCO, LMO, LNMO, etc. The large negative voltage at the anode attracts Lithium ions that pass through the electrolyte and the separator. These Lithium ions finally combine with the high-energy electrons and intercalate into the anode (insert into voids in the crystallographic structure of the carbonaceous anode (graphite) and form Lithium carbide). The corresponding chemical reaction is given in Eq. (2). When all (or majority) of the Lithium ions are moved from the cathode area to the anode area, the battery is said to be charged, and then the following chemical reactions cease.

$$\mathrm{LiMO} \rightarrow \mathrm{MO} + \mathrm{Li}^{+} + \mathrm{e}^{-} \tag{1}$$

$$\mathrm{C}_6 + \mathrm{Li}^{+} + \mathrm{e}^{-} \rightarrow \mathrm{LiC}_6 \tag{2}$$

Even when the charger is removed, Lithium atoms interlaced into the graphite structure of the anode remain there. Nevertheless, as the Lithium ions are combined with the high-energy electrons, they always look for an opportunity to release this energy and pass through the electrolyte to settle in the cathode area. Once an external circuit is formed, in other words, when a load is connected, the necessary condition to achieve this goal is met, and thus the high-energy electrons pass through the load as shown in Fig. 1b. When the electrons pass through the load, they release their energy producing useful work. Chemical reactions related to the discharging process are given in Eqs. (3), (4).

$$\text{At the cathode } \mathrm{MO} + \mathrm{Li}^{+} + \mathrm{e}^{-} \rightarrow \mathrm{LiMO} \tag{3}$$

$$\text{At the anode } \mathrm{LiC}_6 \rightarrow \mathrm{C}_6 + \mathrm{Li}^{+} + \mathrm{e}^{-} \tag{4}$$

As mentioned earlier, charging is the point where EVs interact with the outside world to seek the energy required to operate the vehicle. In this process, charger plays a significant role because it supplies required energy in the form of charging current or in other words, a flow of high-energy electrons. These electrons should come at a suitable level of energy (which is called the charging voltage) and at a suitable rate (which is called the charging current), giving sufficient time for the chemical reactions in Eqs. (1), (2) to occur safely and effectively. Moreover, the flow of electrons should be only in one direction [which is called direct current (DC)].

There are various ways to generate the high-energy electron flow, or simply known as the charging current, while meeting these criteria. Section 6.2 discusses these charging technologies in detail. As the market grows, various products start to come into the EV charging market, and thus, it becomes essential to maintain interoperability among different products. This can be achieved only thorough standardizations. Therefore, Section 6.3 is focused on EV charging standards.

Similar to the fueling stations where traditional vehicles get their source of energy, EVs get their energy from the charging stations that are connected to the power grid. Unlike in fuel stations where the fuel is stored, electrical power is generated in real time, and thus, fluctuations in the EV charging load immediately affect the dynamic stability of the power grid. These issues and potential solutions are discussed in Section 6.4. Overall, this chapter aims to discuss the various charging technologies available for EVs, international standards applied, and impacts of large-scale EV charging on the electrical power grid.

6.2 Charging technologies

Charging technologies applicable for EVs can be broadly classified into three categories namely, conductive charging, inductive charging, and battery swapping. Fig. 2 further illustrates the different types of charging technologies available under these three categories [1]. As the name implies, conductive charging creates a physical connection with wires and metallic contacts [2] for high-energy electrons to flow from charger to the EV battery. This is the simplest, matured, and commonly used technology for charging EV batteries. The scalability and availability at various power levels (Levels 1–3) to suit different EV types make conductive charging to dominate the EV market, including heavy-duty

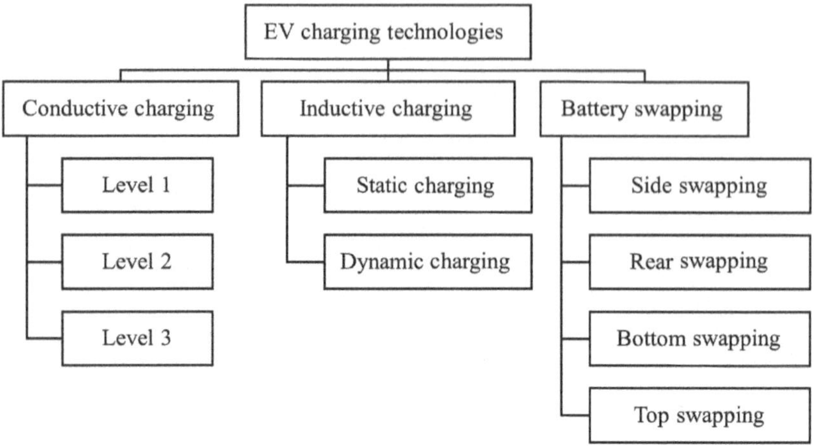

Fig. 2 EV charging technologies.

EVs. The inherent limitation of conductive charging is the need for making a physical contact, which requires the vehicle to be stationary at a specific location. The EV user must plug the charger connector to charge the battery and unplug at the end of charging.

Inductive charging is a solution to this hassle, which eliminates the need for a physical contact between the charger and battery. This is an emerging technology that uses the electromagnetic induction between two coils to transfer energy through air. This can be considered as a loosely coupled transformer where energy flows from the primary coil to the secondary coil through a magnetic field [2]. "Wireless power transfer" is another term used to refer to the same technology, which reflects the fact that there is no physical contact using wires. The absence of a physical contact makes it possible for the EV to get charged while it is moving, which is known as dynamic charging [3]. Even though inductive charging is more flexible, charging power levels are still low compared to conductive charging. Therefore, further developments are required to meet the high-power demands in heavy-duty EVs.

If a heavy-duty EV is operating at a tight schedule with limited time for charging, both the conductive and inductive charging technologies would not be practical. In such a situation, battery swapping is the solution where a charged battery replaces the discharged battery within minutes. Automated battery swapping stations have been developed for this purpose. This subsection explains the abovementioned charging technologies and their applications in heavy-duty EVs.

6.2.1 Conductive charging

A typical conductive charging system is shown in Fig. 3 where the grid power supply acts as the source of electrical energy, and the battery acts as the receiving device. Typically, the power grid provides alternating current (AC) while the battery requires direct current (DC). Owing to this difference in current types, a power converter is required in between to perform the AC to DC conversion and act as the interface. In addition to the AC/DC conversion, the interfacing converter should regulate the voltage and current to match the charging profile of the battery while ensuring the safe operation.

Generally, an EV battery pack consists of a large number of individual battery cells. Thus, if the charging energy is not distributed equally, there is a risk of overcharging particular cell which could lead to premature failure. Therefore, charge balancing is a necessary functionality that should be carried out by the charging setup [4]. Moreover, the thermal stresses and overcurrent protection are other important safety aspects to be taken into account during charging [5]. These functionalities are performed by the battery management system (BMS). Therefore the complete charging setup should come with the charger and the BMS. Depending on the way these two elements are arranged, EV charging configurations can be divided into two categories as onboard chargers and off-board chargers, as shown in Fig. 3.

On-board chargers come as a combination of the power converter and BMS as shown in Fig. 3a [6]. Therefore vehicles with onboard chargers can be directly plugged into the power grid, which is convenient. Nevertheless, the onboard chargers add weight and cost into the EV, and thus there are limitations in increasing the charging power level. The off-board charger, on the other hand, has its charger outside the EV as shown in Fig. 3b. Therefore, in addition to the power cables, there should be communication links as well to share information between the off-board charger and BMS. This approach requires dedicated chargers installed at

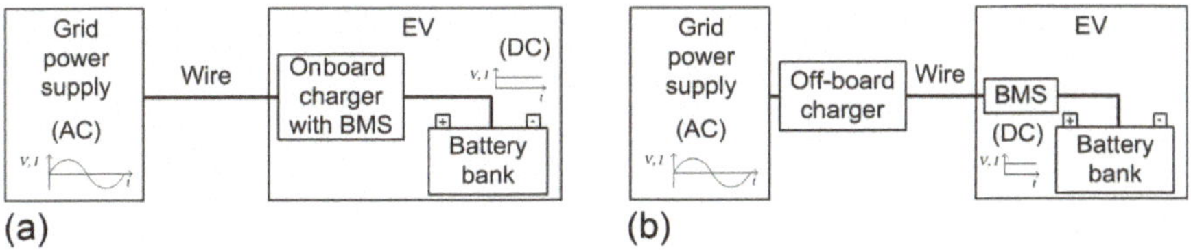

Fig. 3 Typical conductive charging system with an (a) on-board charger and (b) off-board charger.

specific locations to perform charging. As the power converter is situated outside, it can be built to handle increased power levels. Therefore, almost all the high-power EV chargers come as off-board chargers. As this suggests, off-board charging is a suitable approach for heavy-duty EVs.

6.2.1.1 Charging levels

Society of automobile engineers (SAE) has classified EV chargers into three levels as Level 1, Level 2, and Level 3 based on the charging power levels [7]. According to SAEJ1772 standard, typical power levels, operating voltages, and currents for the aforementioned charger categories are summarized in Fig. 4 [1, 8]. On-board charging is predominant in Level 1 charging, while some manufacturers have extended their onboard chargers up to Level 2. AC Level 1 charging is primarily for plugging the EV to the single-phase home power sockets, which can deliver 12–16 A resulting in up to 1.9 kW. This is the slowest charging option, which is suitable for overnight charging. DC Level 1 charging is much more powerful and extends up to 40 kW with voltage levels ranging from 200 to 500 V.

AC Level 2 chargers that operate at 240 V are becoming popular among EV users as they could supply charging power up to 19.2 kW and thereby reduce the charging time.

At higher power levels, the current drawn from the grid could reach up to 80 A, and thus dedicated three-phase connections are required. DC Level 2 chargers come at high charging power levels up to 100 kW. Their operating voltage levels are 200–500 V, and the charging current could reach up to 200 A. They require special charging units to be installed at suitable locations. Level 2 chargers can be installed in garages or places where EVs are parked for a sufficiently long duration. Owing to their

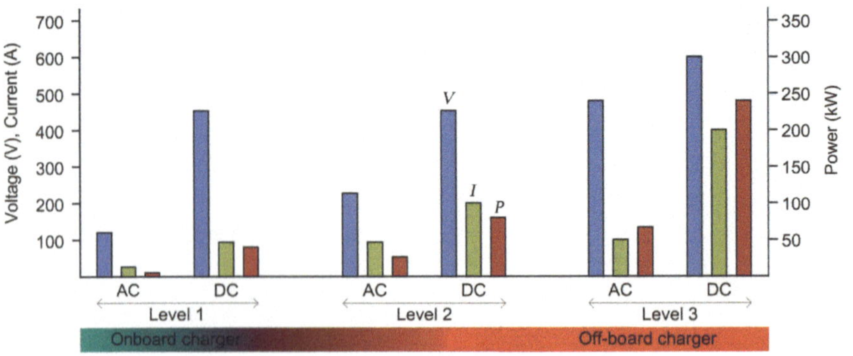

Fig. 4 Charging levels according to SAEJ1772 standard.

high charging power and resultant short charging time, EV charging stations prefer to install Level 2 DC chargers as well. Even though, Level 3 AC chargers are supposed to deliver more power when compared with Level 2 AC chargers, their power levels are lower than DC chargers and thus not becoming attractive [1]. Instead, technologies are being developed to increase the DC charging power levels beyond 100 kW. The Level 3 DC fast charger is the result of these advancements, which could deliver up to 240 kW of charging power from 200 to 600 V supply. More advanced DC fast chargers are coming into the market, which could deliver more than 400 kW. These are the types of chargers required for charging heavy-duty EVs.

6.2.2 Inductive charging

The risk of electrocution and other hassles with cables and connectors can be eliminated with wireless charging. As shown in Fig. 5, there are various types of wireless power transfer technologies, of which inductive coupling-based wireless power transfer is the feasible technology for charging EV batteries. This technology is generally known as inductive power transfer (IPT) [1]. IPT-based EV charging systems use two coils namely, primary coil and secondary coil. The primary coil is often buried at the charging ground, and the secondary coil is installed at the bottom of the vehicle [9]. When the two coils are aligned and the primary coil is energized, a varying magnetic field is estabilished and linked with the secondary coil to induces an electromotive force (emf) in it. Fig. 6a shows the graphical representation of an IPT system with the primary and secondary coils.

According to the Ampere's Law, current in the primary coil, I_p (A), produces magnetic flux. Some of these flux link with the secondary coil,

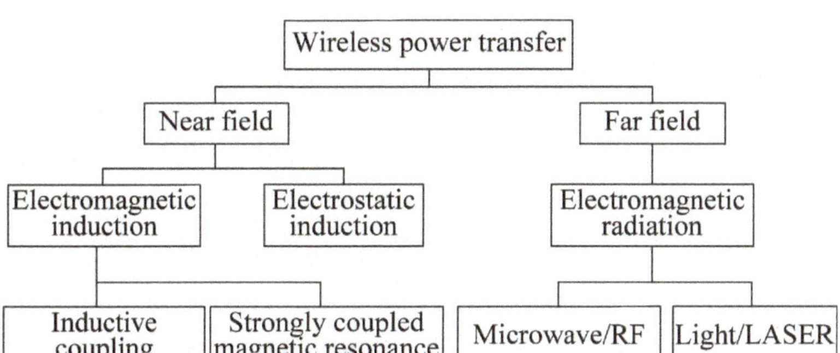

Fig. 5 Wireless power transfer methods.

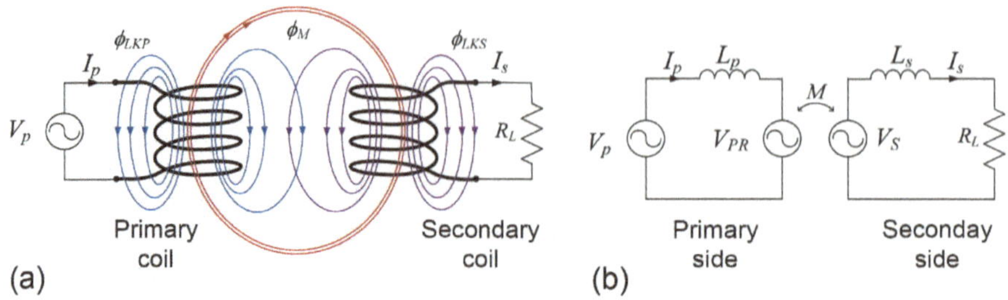

Fig. 6 (a) An IPT system with associated magnetic flux and (b) equivalent circuit.

which is known as mutual flux, and is marked as ϕ_M (Wb) in Fig. 6a. According to the Faraday's Law, this flux induces an emf in the secondary coil enabling power to be transferred to the load R_L (Ω). Owing to the air gap and resultant weak coupling between the two coils, a significant amount of flux produced by the primary coil leaks into the surrounding. This leakage flux at the primary side is marked as ϕ_{LKP}. When there is a current flowing in the secondary coil, it also produces magnetic flux. Part of this magnetic flux leaks into the surrounding of the secondary coil, which is marked as ϕ_{LKS} (Wb) in Fig. 6a. The rest of the flux produced in the secondary coil links with the primary coil. The abovementioned mutual flux accounts for this magnetic flux coming from the secondary coil as well.

In circuit analysis, leakage flux can be considered as self-inductance. Therefore the equivalent electrical circuit shown in Fig. 6b contains two self-inductances L_p and L_s (H) to account for the leakage flux. The mutual flux is represented as mutual inductance $M(H)$. The equivalent circuit shown in Fig. 6b resembles a transformer. Nevertheless, the IPT systems differ from standard transformers in terms of the degree of coupling. In IPT systems, the separation between primary and secondary coils is large, and thus the reluctance of the magnetic flux path is high. Therefore the coils are said to be loosely coupled as opposed to strongly coupled coils in transformers. As mentioned earlier, this leakage flux increases the self-inductance in both primary and secondary coils. The drawback of self-inductances is that they significantly limit the power transfer capability. This is similar to the reactive power in AC power systems, and in fact, the power associated with leakage flux is called reactive power in IPT systems. Similar to the power factor correction in AC power systems, the self-inductances in primary and secondary coils can be compensated by adding capacitors as shown in Fig. 7. With proper sizing, the two capacitors, C_p and C_s, can compensate the two self-inductances L_p and L_s.

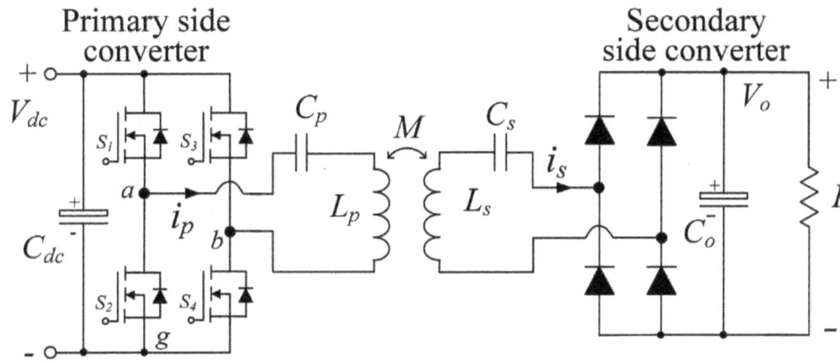

Fig. 7 Series compensation in an IPT system.

Emerging applications, such as vehicle-to-grid (V2G) systems, require bidirectional power flow in IPT systems for exchanging power with the grid. This requires replacement of the passive rectifier at the secondary side with an active rectifier. The resultant converter arrangement shown in Fig. 8a is known as the dual active bridge (DAB). Since the two converters can be controlled independently, they can be considered as two independent active AC sources. The angles ϕ_1 and ϕ_2 shown in Fig. 8b are the phase shifts between the two legs of each converter. These angles represent the duty cycle of each converter output voltage. The angle θ is the phase shift between the two voltage waveforms produced by the DAB converter system. Power associated with each converter can be controlled by controlling the angles ϕ_1 and ϕ_2, and the direction of power flow can be controlled by controlling the angle θ. In the circuit shown in Fig. 8a, the combination of an inductor and a capacitor is used as the compensation network in both the primary and secondary side. More information on compensation techniques can be found at Ref. [10].

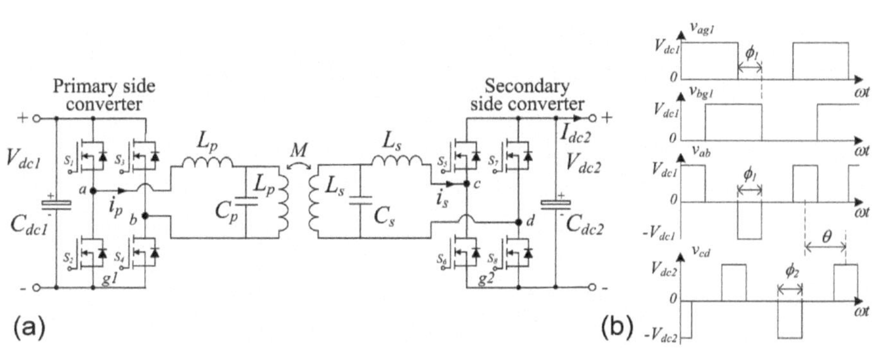

Fig. 8 Bidirectional wireless power transfer system (a) schematic diagram and (b) waveforms.

6.2.2.1 Static inductive charging

As the name suggests, static charging refers to the charging of the vehicle when it is parked. The vehicle should be parked in a way that the two coils are aligned correctly to achieve maximum power transfer efficiency. Recent developments in IPT technology allow a certain degree of misalignments making it more feasible in practical implementations [11]. The air gap between the two coils is another factor that affects the amount of power that could be transferred and the efficiency. The current state of technology allows up to 100 kW to be transferred with an air gap of 125 mm [12]. This type of charging is suitable for home garages, shopping malls, and car parks, where the vehicle is being parked for sufficiently long time.

6.2.2.2 Dynamic inductive charging

Dynamic charging refers to the charging of EV battery while the vehicle is moving. Owing to its numerous advantages, dynamic charging is gaining increased attention in the EV industry [3]. Several initiatives have been reported worldwide to develop special tracks for dynamic charging. This enables EV manufacturers to reduce the size of onboard storage unit because it can be charged while moving [2]. Nevertheless, the cost associated with installations is the major challenge in implementing dynamic charging.

6.2.3 Battery swapping

Conductive or inductive charging methods mentioned earlier may not be suitable for heavy-duty EVs operating in tight schedules. These vehicles require their batteries to be charged within a short period, and thus very high charging levels are required. Inductive charging technologies are not matured yet to meet such high-power levels. Even though conductive charging could handle very high charging power levels with DC fast charging, batteries get overstressed and result in performance degradation and reduction in cycle life. The promising solution for such applications is battery swapping, which involves the replacement of depleted battery with a charged battery [1]. Battery swapping stations have been built with robotic arms to automate the swapping process and thereby reduce time required for swapping to several minutes. Another advantage of battery swapping is the possibility of using the batteries at swapping stations as grid support units where batteries could supply power to the grid during peak times and charge during off-peak periods. Presently, there are some challenges in implementing battery swapping, which includes infrastructure, interchangeability, feasibility, and ownership of the battery

[6]. Swapping stations require very complex and expansive infrastructure when compared with conductive or inductive charging. Each swapping station requires a significantly large storage area, which might not be possible to find in certain areas. Moreover, battery swapping stations present a massive demand on the grid and may require installing new distribution infrastructure, including transformers. Different manufacturers have their own battery system designs, which might lead to interoperability and feasibility issues of battery swapping. The ownership of the battery under swapping arrangement is another issue, which may end up as an additional burden to the swapping station. Nevertheless, various business models are presently being developed to overcome these limitations and promote battery swapping. More information on battery swapping can be found in Chapter 5.

6.3 Charging standards

With the rapid growth in economic potential of EVs, significantly large number of manufacturers came up with new designs for EVs and charging solutions. EVs themselves are free to have proprietary designs and features to gain competitive advantage and become leaders in the market. Nevertheless, charging is the most critical aspect where all these different products interact with each other, and therefore it is natural to demand interoperability among different EVs and charging facilities. As a solution, worldwide regulatory bodies have established standards for EV charging. These standards can be divided into three main groups namely, EV charging component standards, EV grid integration standards, and safety standards [1]. Fig. 9 shows a graphical illustration of these three groups. The acronyms are defined as follows: SAE: Society of Automotive Engineers, ISO: International Organization for Standardization, IEEE: Institute of Electrical and Electronics Engineers (IEEE), UL: Underwriters' Laboratories, NFPA: National Fire Protection Association, NEC: National Electric Code, NEMA: National Electrical Manufacturers Association, ANSI: American National Standards Institute, GB: Guobiao, and IEC: International Electrotechnical Commission.

The SAE and IEEE standards are mainly used in the United States. The IEC develops standards for electrical, electronic, and other related technologies and is predominantly used in Europe. The CHAdeMO standards are used in Japan. China uses the Guobiao (GB/T) standard issued by the Standardization Administration of China and the Chinese National Committee of ISO and IEC.

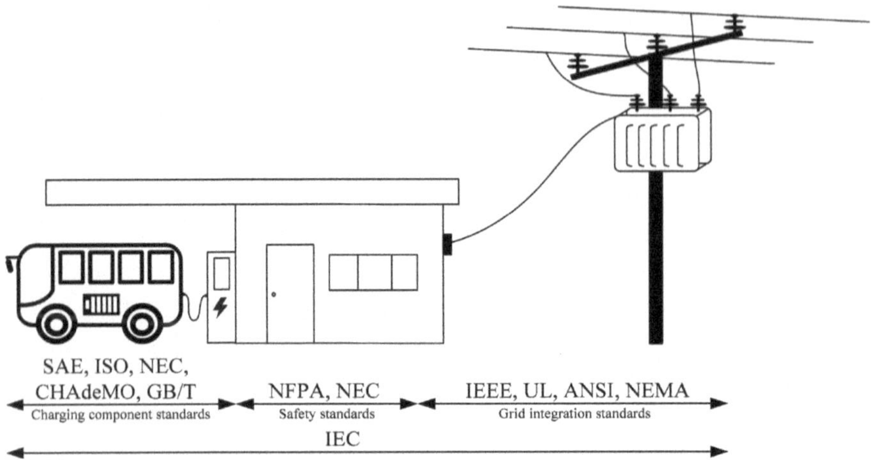

Fig. 9 EV charging standards and application areas.

The SAEJ2293 standard has two sections, namely, J2293-1 and J2293-2, which collectively establish requirements for off-board charging equipment. J2293-1 stipulates the power requirements and system architecture for conductive (AC and DC) charging and inductive charging. Communication requirements and network architecture for EV charging are discussed in J2293-2. Equipment ratings, including circuit breaker current and charging voltage, are stipulated in SAEJ1772 for all the levels and types of EV charging. SAEJ1773 specifies the requirements for inductive charging schemes. SAEJ2847 and SAEJ2836, along with SAEJ1772, specify requirements to be fulfilled by the communication link between EVs and charging stations. SAEJ2954 standard specifies wireless charging. This covers payment establishment and autonomous charging as well.

NFPA standards cover wiring and safety at the use side of the electricity supply. They include electric conductors and equipment installed within buildings and other structures.

IEEE1547 is the standard for interconnecting distributed resources with electric power systems, which stipulates requirements on the performance, operation, testing, safety, and maintenance. UL1741 standard discusses protection of the power conversion equipment and their grid connection.

As depicted in Fig. 9, IEC61851 covers overall the operation of EV conductive charging up to 1000V AC and 1500V DC. IEC 61980 provides a standard for inductive charging up to 1000V AC and 1500V DC. The standard for plugs, socket-outlets, and vehicle connectors are stipulated in IEC62196. More information and discussion on other relevant standards can be found in Ref. [1].

6.4 Effects of fast charging on EV batteries

From the user point of view, time required for charging the battery for the next journey is a crucial factor, and thus DC fast chargers, which effectively reduce the charging time, are becoming popular. Even though fast charging reduces the charging time, as shown in Fig. 10, it stresses the battery at various levels in various forms [13]. This is mainly because of pushing the battery to extremes and forcing the reactions in Eqs. (1), (2) mentioned above to occur at a very high rate. As a result, at the atomic level, thermodynamic properties, lattice stability transfer of charges, and diffusion of charges get affected. At the molecular level, Lithium ion plating and solid electrolyte interphase (SEI) growth could occur as a result of fast charging. Lithium plating refers to the formation of metallic Lithium instead of intercalating Lithium ions into the anode. Deposited Lithium ions are no longer available for the discharging process, and thus battery performance degrades. This could even lead to short-circuit at extremes. SEI formation and growth on electrode surfaces from decomposed products of electrolytes occurs consuming active Lithium and electrolyte materials, leading to capacity fading, increasing battery resistance, and low power density. Moreover, cracks could occur leading to mechanical failures inside the battery cell.

As a result of reactions occurring at a faster rate, more heat gets generated within the cell [5, 14]. The overall heat of the cell (Q) is the result of reversible heat (Q_{rev}), irreversible heat (Q_{irr}), and heat generated at the tabs (Q_{tab}) as follows:

$$Q = Q_{rev} + Q_{irr} + Q_{tab} \tag{5}$$

The other forms of heat generation, such as of mixing and phase transition, are neglected here for the simplicity. More information of

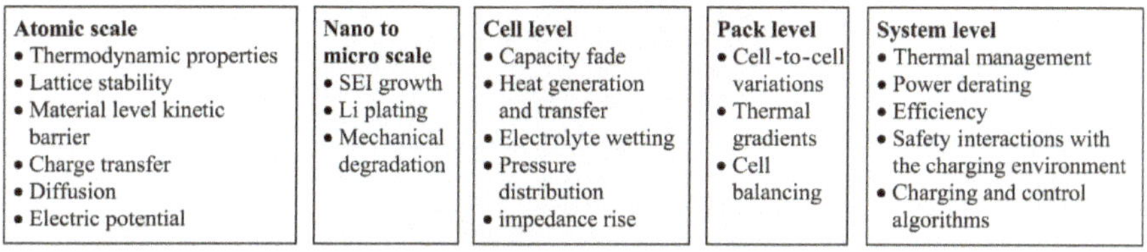

Fig. 10 Effects of fast charging at different scales of EV batteries.

these mechanisms can be found at Ref. [15]. The reversible heat of the electrodes is owing to entropy change and can be either positive or negative. The irreversible heat (Q_{irr}) is attributed to the total heat generated in the electrodes, separator, and current collectors as a result of the electrochemical reaction and ohmic (electronic and ionic) potential drop. In general, Q_{irr} includes both reaction heat (Q_r) and ohmic heat (Q_{ohmic}). The ohmic heat itself consists of ionic ($Q_{ohmic,ionic}$) and electronic ($Q_{ohmic,\ elec}$).

$$Q_{irr} = Q_r + Q_{ohmic,\,ionic} + Q_{ohmic,\,elec} \tag{6}$$

The generated heat in the tabs is mainly owing to ohmic resistance and change of cross-sectional area in the tab which is in contact with the current collector.

$$Q_{tab} = I^2(R + R_C) \tag{7}$$

where I is the battery cell current, R resistance of the current collector, and R_C is the electrical contact resistance between the current collecting tab and the lead wire.

The generation of heat varies from point to point of the cell. Depending on the way the generated heat is taken out from the cell, the temperature profile of the cell varies. This escalates when the cells are packed to form battery modules. In extreme cases, thermal hotspot could be generated, leading to premature failures. Fig. 11 shows the uneven temperature

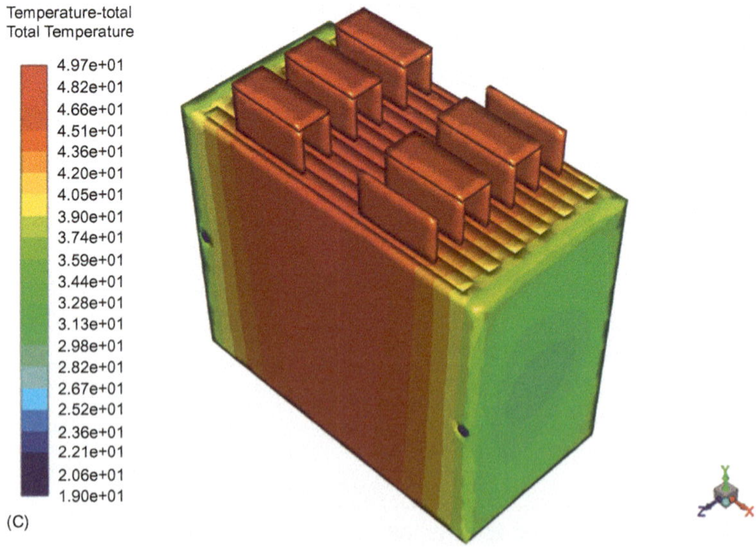

Fig. 11 Temperature distribution of a six-cell battery pack with liquid cooling [5].

distribution in a six-cell battery pack with liquid cooling plates at the sides [5]. The area between the terminals shows the highest temperature. This area gets overstressed owing to fast charging and could lead to premature failures. Therefore thermal management and derating of charging power are essential in fast charging to save battery life. Therefore fast charging requires advanced battery management techniques and complex charging algorithms.

6.5 Grid impact of EV charging

From the charging point of view, EV systems face a significant challenge in managing the electrical power supply and demand in real time, which is not present in traditional fossil fuel-based vehicles. Fossil fuels are generally kept stored in tanks, and thus there is a buffer to decouple the dynamics of the demand and the supply. Even if there is a sudden change in the demand or supply or both, it takes some time to create an imbalance, and thus appropriate actions can be taken to restore the balance. Such flexibility does not exist in EV charging because the power required to charge EV batteries is generated and delivered in real time. Therefore if an EV started to draw a significantly large amount of power from the grid, it creates a momentary imbalance between the supply and demand. In uncontrolled charging, EV receives power immediately from the grid when connected and starts charging rapidly. Once charging has started, it usually continues until the battery is fully charged or the user decides to stop charging. Even though this method is simple and straightforward, it directly exposes the grid because the grid operators do not receive information about the loads [16]. Because the generator controller is not aware of the change, power generation remains unchanged, leading to an imbalance in the demand and supply. As a result, the generator starts to slow down and the grid frequency drops. Control system of the generator detects the deviation and gradually increase the power generation to restore the frequency. Depending on the load added to the grid, the recovery time and frequency dip vary, and in extreme cases, they can exceed the limits resulting in grid instability [17]. Generally, electrical power grids have significantly large inertia, and thus adding small EV charging loads do not lead to instabilities. Nevertheless, if large EV charging loads or a large number of small EV charging loads are connected, the grid might become unstable [18]. Therefore uncontrolled charging is not suitable, especially for heavy-duty EVs. As a solution, several controlled charging methods have been developed, as shown in Fig. 12 [19].

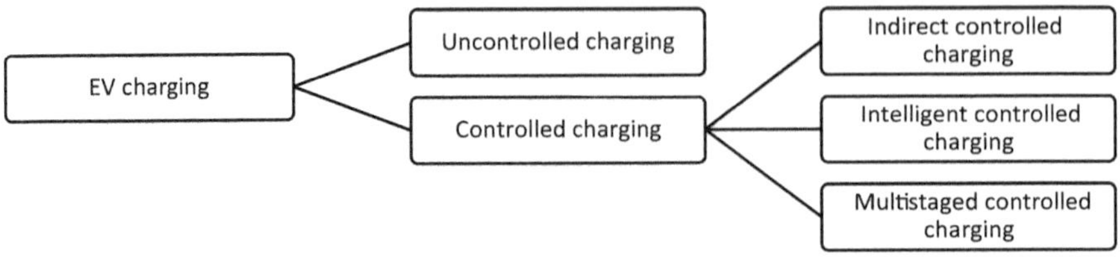

Fig. 12 EV charging coordination techniques.

Controlled EV charging has recently become popular among the industry and academia, resulting in a large number of techniques developed to gather vehicle charging needs, coordinate EVs with charging stations, and communicate with the grid operators in advance to avoid potential imbalances. Moreover, the possibility of optimizing the objectives of EV users, charging infrastructure owners and grid operators is another advantage of controlled charging. As shown in Fig. 12, the three main categories of controlled charging are indirect controlled charging, intelligent controlled charging, and multistage controlled charging.

As the name implies, indirect controlled charging does not constrain charging parameters, such as charging power and duration. Instead, it considers user behavior and decisions to control out-of-system parameters that influence the charging method. The indirect control can be broadly classified into two categories namely, spatial load shifting and temporal load shifting. In spatial load shifting, users are encouraged with financial incentives to use less congested charging stations and thereby distribute the load. Temporal load shifting encourages users to charge EVs during off-peak hours with a reduced cost of electricity and thereby prevent grid overload [19].

Intelligent control is a data-driven approach where the coordination mechanism searches for combinations that optimize the use of existing resources within given constrains. Data collection and communication play a crucial role to link EVs, charging stations, and transmission/distribution system operators. The objectives of optimization could be charging power, cost, or efficiency [20]. The constrains are often found to be initial demand for energy, battery capacity, number of generation units, network structure, efficiency of transformers, voltage, frequency, and other power quality requirements.

Multistage hierarchical controlled charging method consists of a priority-based multilevel decision-making tool. It provides a unique solution using decision-based control through a genetic algorithm (GA) with fuzzy- or artificial intelligence-based control tools [19]. The four main areas considered in this method are the load capacity of the infrastructure, priorities of the EVs, such as battery state of charge (SOC), charging cost, and time-of-use. This considers bidirection power flow and allowing EVs to inject power to the grid during peak times to achieve frequency regulation meet power quality requirements. When a mismatch between the supply and demand is detected, the central controller communicates with the charging stations to determine the maximum number of EVs that can be used to inject power to the grid or absorb power from the grid. This is presently in the development stage and would be implemented in the future [16, 19]. Table 1 summarizes controlled and uncontrolled charging methods (Table 1).

Table 1 Comparison of controlled and uncontrolled charging methods.

Charging type	Advantages	Disadvantages	Effects on the grid
Uncontrolled charging	Easy to use	Grid could be overloaded	Power quality degradation
	Vehicle starts charging as soon as plugged in	Large voltage fluctuations could occur	Overloading of grid elements
		High cost of electricity	High power losses
		Low power factor	High electricity cost
			Increase in uncertainties
Controlled charging	Peak charging power is reduced	Distribution network design becomes complex	Peak load is managed easily
	Demand profile is smoothened	Battery SOC could be reduced fast	Easy coordination between grid and load
	Vehicle to grid power transfer is possible	There could be more losses incur in the grid	Increased reliability and stability in power grid
	Distribution network and resources are effectively and optimally used	There could instances where charging is unavailable	Reduced burden on grid elements
		Different controlled techniques are required in different critical situations	

6.6 Case study

The return trip of an electric bus operating in an urban area was taken as a case study to find out charging system requirements. Graphical representation of the electric bus model used in this study is shown in Fig. 13 [21]. The corresponding system parameters are given in Table 2. The speed profile shown in Fig. 14a was derived from the NEDC (New European Driving Cycle) for a return trip with an upper limit of 80 km/h. The corresponding traction power is shown in Fig. 14b for the two cases, with and without regenerative braking. The case without regenerative braking shows only positive traction power. The case with regenerative power shows negative power as well. Nevertheless, the negative power is capped at 75 kW, which is 50% of the peak power owing to charging limitations of the battery bank. Moreover, the lower speed limit for regeneration action is set to 30 km/h in this study. The amount of energy consumed at different times is plotted in Fig. 14c, where at the end of the journey, the total energy consumed for traction becomes 10.2 kWh for the case without regenerative braking. With the regenerative braking, the required amount of energy drops to 9 kWh, which is a significant saving in the long run. When the other loads such as air conditioning and losses are considered, the total energy required to complete the 40-min journey can be safely assumed to be 15 kWh. If the charging time available before

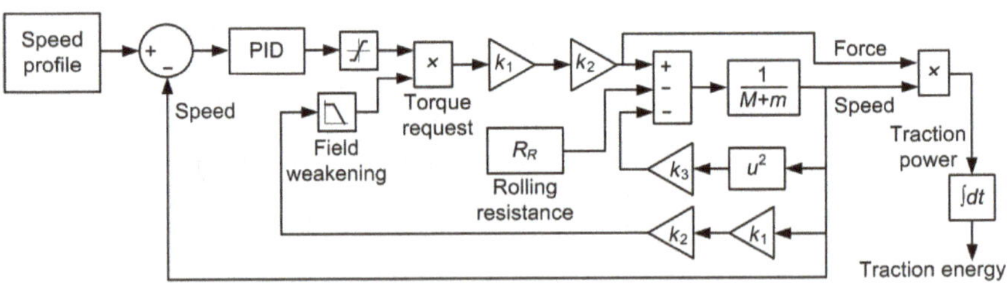

Fig. 13 Graphical representation of the electric bus model.

Table 2 System parameters of the electric bus simulation model.

Parameter	Value	Parameter	Value
Gear ration, k_1	8	Vehicle body mass, M	10,000 kg
Tyre radius inverse, k_2	1/0.52	Passenger mass, m	1200 kg
Aerodynamic drag, k_3	$1.225 \times 0.7 \times 4$	Travel duration	40 min
Rolling resistance, R_R	$0.02 \times 1000 \times 9.81$	Travel distance	20 km

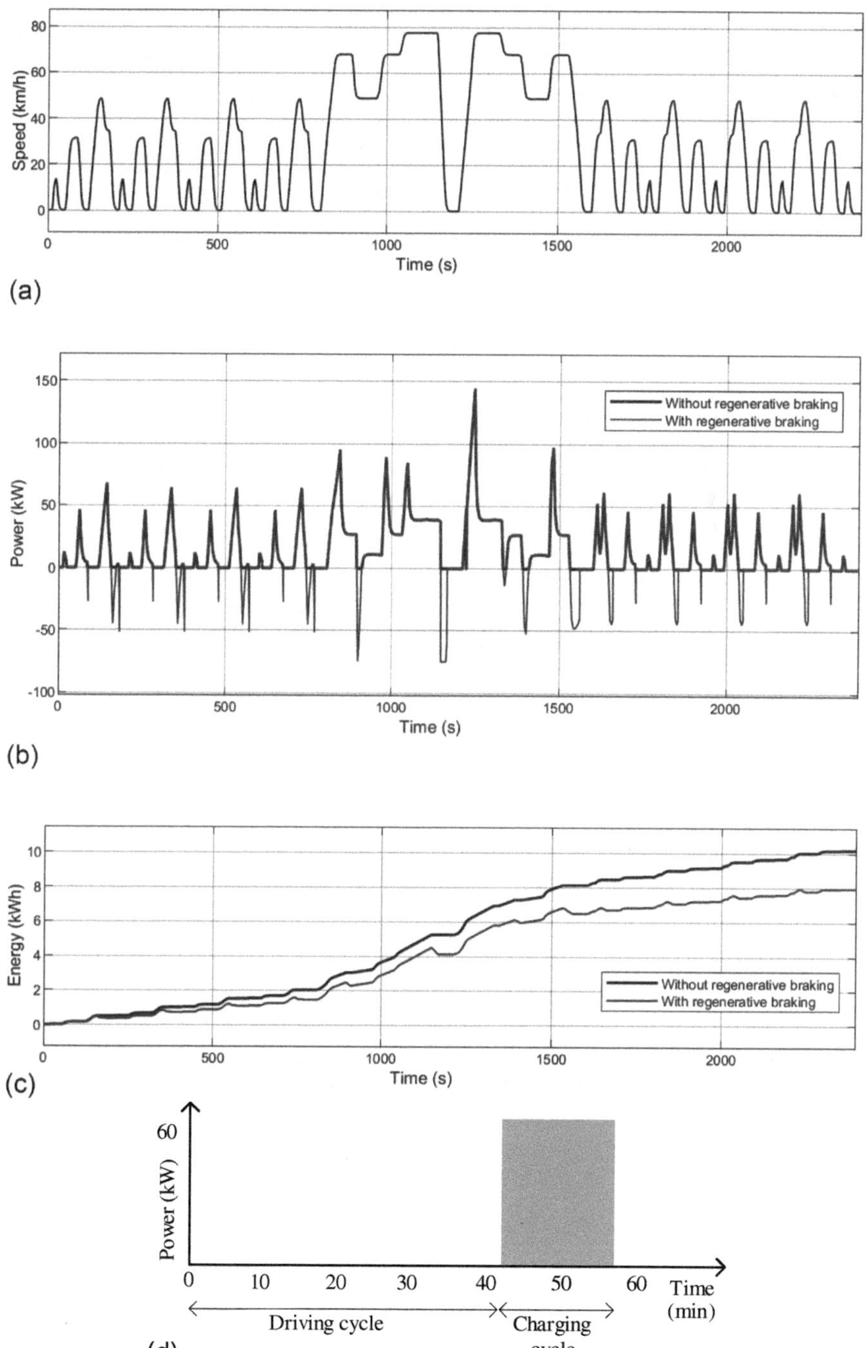

Fig. 14 Case study results (a) speed profile, (b) traction power, (c) traction energy, and (d) charging cycle.

starting the next journey is 15 min, the charger should be able to provide 15 kWh within this period. As shown in Fig. 14d, if uniform charging is assumed, this brings the charging power requirement to 60 kW. Therefore according to Fig. 4, a Level 2 off-board charging facility is required for this electric bus. However, it should be noted that this is only a rough calculation. More information on battery sizing can be found at Refs. [22,23].

6.7 Summary

This chapter aimed to provide readers with an understanding of the charging and discharging process and state-of-the-art charging technologies. Conductive charging is introduced as the most common and technically feasible technology for charging heavy-duty EVs. Nevertheless, the potential of static and dynamic wireless charging has also been discussed as an emerging technology. Charging levels and standards applied in different countries are also introduced, highlighting that fast charging is becoming a trend in the EV market. Negative impacts of fast charging on battery health are also elaborated with emphasis on hotspots. Besides, the effects of fast charging and increased charging load on the grid are discussed with potential solutions with coordinated charging. Finally, a case study is presented to estimate the charging power requirement of an electric bus.

References

[1] A. Ahmad, et al., A review of the electric vehicle charging techniques, standards, progression and evolution of EV technologies in Germany, Smart Sci. 6 (1) (2018) 36–53.

[2] C. Panchal, S. Stegen, J. Lu, Review of static and dynamic wireless electric vehicle charging system, Eng. Sci. Technol. 21 (5) (2018) 922–937.

[3] K.A. Kalwar, M. Aamir, S. Mekhilef, Inductively coupled power transfer (ICPT) for electric vehicle charging—a review, Renew. Sust. Energ. Rev. 47 (2015) 462–475.

[4] Z.B. Omariba, L. Zhang, D. Sun, Review of battery cell balancing methodologies for optimizing battery pack performance in electric vehicles, IEEE Access 7 (2019) 129335–129352.

[5] E. Alston, S. Jayasinghe, C. Baguley, U. Madawala, Thermal management of an electric ferry lithium-ion battery system. in: 2018 IEEE 4th Southern Power Electronics Conference (SPEC), Singapore, 2018, pp. 1–4, https://doi.org/10.1109/SPEC.2018.8635922.

[6] A.S. Al-Ogaili, et al., Review on scheduling, clustering, and forecasting strategies for controlling electric vehicle charging: challenges and recommendations, IEEE Access 7 (2019) 128353–128371.

[7] H.S. Das, et al., Electric vehicles standards, charging infrastructure, and impact on grid integration: a technological review, Renew. Sust. Energ. Rev. 120 (2020) 109618.
[8] I. Rahman, et al., Review of recent trends in optimization techniques for plug-in hybrid, and electric vehicle charging infrastructures, Renew. Sust. Energ. Rev. 58 (2016) 1039–1047.
[9] J.S. Gill, et al., Infrastructure cost issues related to inductively coupled power transfer for electric vehicles, Procedia Comput. Sci. 32 (2014) 545–552.
[10] W. Zhang, C.C. Mi, Compensation topologies of high-power wireless power transfer systems, IEEE Trans. Veh. Technol. 65 (6) (2016) 4768–4778.
[11] L. Zhao, et al., A misalignment-tolerant series-hybrid wireless EV charging system with integrated magnetics, IEEE Trans. Power Electron. 34 (2) (2019) 1276–1285.
[12] V.P. Galigekere, et al., Design and implementation of an optimized 100 kW stationary wireless charging system for EV battery recharging. in: 2018 IEEE Energy Conversion Congress and Exposition (ECCE), Portland, OR, 2018, pp. 3587–3592, https://doi.org/10.1109/ECCE.2018.8557590.
[13] A. Tomaszewska, et al., Lithium-ion battery fast charging: a review, eTransportation 1 (2019) 100011.
[14] C. Chen, F. Shang, M. Salameh, M. Krishnamurthy, Challenges and advancements in fast charging solutions for EVs: a technological review. in: 2018 IEEE Transportation Electrification Conference and Expo (ITEC), Long Beach, CA, 2018, pp. 695–701, https://doi.org/10.1109/ITEC.2018.8450139.
[15] S. Arora, A. Kapoor, Experimental study of heat generation rate during discharge of LiFePO4 pouch cells of different nominal capacities and thickness, Batteries 5 (4) (2019) 70.
[16] Y. Zheng, et al., Integrating plug-in electric vehicles into power grids: a comprehensive review on power interaction mode, scheduling methodology and mathematical foundation, Renew. Sust. Energ. Rev. 112 (2019) 424–439.
[17] Y. Huang, K.M. Kockelman, Electric vehicle charging station locations: elastic demand, station congestion, and network equilibrium, Transp. Res. Part D: Transp. Environ. 78 (2020) 102179.
[18] A.G. Anastasiadis, et al., Effects of increased electric vehicles into a distribution network, Energy Procedia 157 (2019) 586–593.
[19] T.U. Solanke, et al., A review of strategic charging–discharging control of grid-connected electric vehicles, J. Energy Storage 28 (2020) 101193.
[20] D. Wu, N. Radhakrishnan, S. Huang, A hierarchical charging control of plug-in electric vehicles with simple flexibility model, Appl. Energy 253 (2019) 113490.
[21] J. Vepsäläinen, K. Otto, A. Lajunen, K. Tammi, Computationally efficient model for energy demand prediction of electric city bus in varying operating conditions, Energy 169C (2019) 433–443.
[22] Z. Gao, et al., Battery capacity and recharging needs for electric buses in city transit service, Energy 122 (2017) 588–600.
[23] O. Teichert, et al., Joint optimization of vehicle battery pack capacity and charging infrastructure for electrified public bus systems, IEEE Trans. Transp. Electr. 5 (3) (2019) 672–682.

PART III

Control Systems

CHAPTER 7

Drivetrain Control System in Heavy-duty Electric Vehicle Applications

7.1 Introduction

Various vehicle operation modes are discussed in Chapter 9. This chapter discusses the required control system parameters that need to be considered for optimal control of the electric vehicle (EV) drivetrain in heavy-duty vehicles. These strategies include torque mapping in driving and regenerative braking conditions and zero (limiting) torque conditions that need to be maintained. This chapter also covers the required parameter settings in the inverter drives according to the drivetrain motor type and specifications.

Drive mode is defined when the vehicle is ready to drive. It means that the driver has requested to drive the car (e.g., turn the Ignition Key to the ON position). In an EV, the supervisory controller (SC) receives the driver's inputs and proceeds as follows:

- Check and confirm whether the battery status is OK and there is no fault.
- Turn on the motor controller and start the controller area network (CAN) communication.
- Connect the high voltage (HV) to the motor controller through the precharge process.
- Check and confirm the motor controller status is OK and there is no fault.
- Enable the motor controller and inform the driver that the vehicle is ready to drive.

Ready to drive state practically means the driver can control the delivered torque to the wheels through the accelerator pedal, brake pedal, and gear position buttons.

7.2 Drivetrain torque control in heavy-duty electric vehicles

Controlling the drivetrain torque is the most critical item that defines the vehicle operation on-road and influences the driving experience for the

Heavy-duty Electric Vehicles
https://doi.org/10.1016/B978-0-12-818126-3.00001-4

driver and passengers. Moreover, drivetrain torque mapping highly depends on the application requirements. However, it is also possible to have multiple driving modes that can be selected by the driver for different road conditions. For instance, many passenger cars have an eco-drive mode for city traffic these days. In summary, Ready to drive state in EV SC systems includes the following major suboperation modes:

- **Drive mode**: It refers to the state that the electric motor is delivering torque to the wheels according to the accelerator pedal position ($\tau \times \omega > 0$, where τ is torque and ω is speed).
- **Brake mode**: It is the condition when the brake pedal is pressed. This operation mode can be divided into two submodes:
  - **Zero torque command**: No regenerative braking is possible. Therefore the torque command to the motor controller must be zero.
  - **Regenerative braking**: It refers to the state where the electric motor is generating power from the wheels according to the brake pedal position ($\tau \times \omega < 0$, where τ is torque and ω is speed).
- **Torque limiting mode**: It involves the various scenarios where the drivetrain torque must be limited or becomes zero in certain fault conditions. These scenarios are discussed in fault-handling strategies in Chapter 9.

7.2.1 Torque mapping strategies in various modes

Torque mapping strategies in EVs are mainly based on the application requirements. Torque mapping in a high-performance passenger car is different from an electric city bus where the former demands a high acceleration and speed operation and the latter needs smooth and average speed operation. Various vehicle torque control strategies are reported in the literature with vehicle stability control and optimal energy consumption as the main objectives [1]. Energy consumption efficiency is the most important aspect of torque mapping in heavy-duty electric vehicles mainly owing to the high power requirements and the presence of large onboard energy resources. Therefore optimization of the battery energy consumption leads to increased mileage. A survey shows that 80% of bus drivers prefer to charge the battery when the SoC is about 40% rather than continue driving [2]. In addition to the above conditions, smooth drive experience for the driver and the passengers is also significant. In addition to all the aspects mentioned earlier, we need to consider the torque limiting strategies in different conditions to maintain safety and optimal performance of the EV drivetrain.

7.2.1.1 Drive mode torque mapping

In Drive Mode, the electric motor(s) delivers power to the wheels in either direction. The main driver inputs are accelerator pedal, brake pedal, and gear position. The brake pedal has higher priority when compared with the accelerator pedal. Therefore if both pedals are pressed at the same time, then the accelerator pedal value will be overridden to zero. Accelerator pedals have dual potentiometers that one follows the other. Based on the position of the pedal, the SC translates the analog voltage from the potentiometer to 0%–100% after filtering the signal; where 0% means the pedal is not pressed, and 100% means the pedal is fully pressed. The Brake pedals in heavy-duty applications usually have ON/OFF type switches to indicate the brake pedal is pressed or not; and in some cases, they have pulse width modulation (PWM) signal output that its duty cycle shows the percentage of the pressed brake pedal and can be translated to 0%–100% by the SC.

In heavy-duty vehicles, change of the gear between reverse, neutral, and drive positions is possible only if the vehicle speed is below a certain limit and in some cases, only if the brake pedal is pressed. This can be implemented mechanically or in SC software. Gear position drive and reverse allows torque commands to the motor controller, while park and neutral positions override the torque command to zero. Park gear in the heavy-duty vehicles directly actuates the air pressure in the brake lines same as the brake pedal.

The accelerator and brake pedal values get mapped to the torque command to the motor. Brake pedal mapping will be explained in the next section. Fig. 1 shows the torque mapping examples based on the accelerator pedal values. The mapping output is a percentage of the maximum available torque of the electric motor.

The maximum torque that the electric motor can deliver varies depending on the motor RPM and needs to be extracted from the electric motor datasheet. Fig. 2 shows a typical torque versus speed characteristics of an EV electric motor with a rated speed of 2400 RPM and a maximum speed of 4000 RPM. Therefore in reality, the commanded torque value is a function of driver inputs and electric motor actual RPM.

Linear torque mapping is the simplest method to translate the accelerator pedal position to the commanded torque directly. The actual acceleration of the vehicle depends on the vehicle mass and delivered torque to the wheels. In heavy-duty vehicle applications, low acceleration mapping is more common owing to the high inertia of the vehicle. Another critical parameter in vehicle drivetrain control is the rate of change of torque

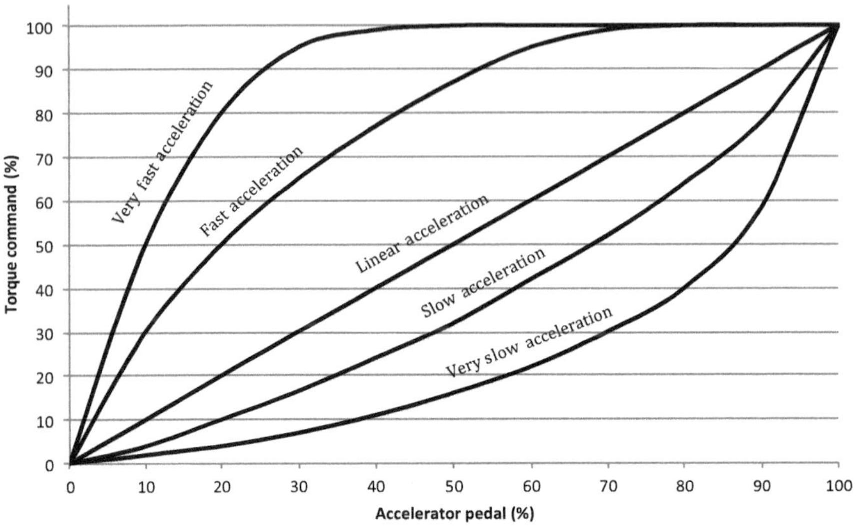

Fig. 1 Possible torque mapping based on the accelerator pedal.

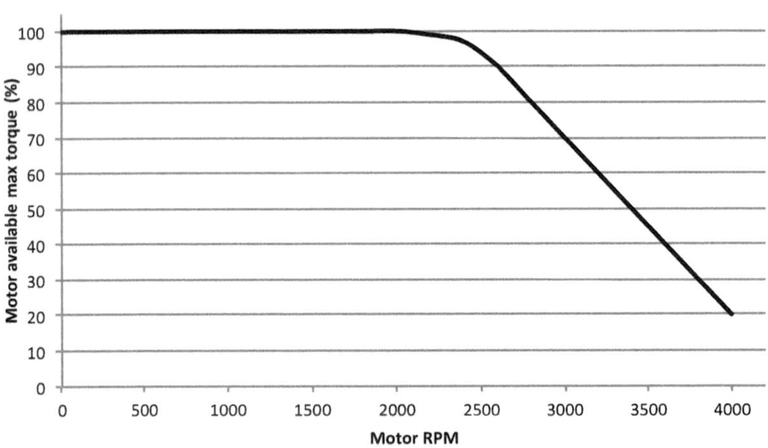

Fig. 2 A typical torque versus speed characteristics of an EV electric motor.

command to the motor controller that must be controlled based on the vehicle speed. Sudden changes of torque demand by the driver through the accelerator pedal must be delivered to the wheel with a predefined acceleration and deceleration rate according to the vehicle speed to maintain the smooth speed change of the vehicle. The final tuning of the torque-mapping table is possible after doing multiple tests and data analysis according to the application requirements.

Demanding high torque for an extended period at low speed could result in damaging the motor winding owing to the heat generated by the high current. Electric motors can deliver the peak torque to the load for a short time (usually less than 60 s); therefore it is essential to select a motor whose peak torque value is higher than continuous peak torque demand in respective applications.

Multiple different methods have been reported in the literature for improving the energy efficiency of EVs [3–6]. Energy efficiency is not in the scope of this chapter. Nevertheless, it is good to highlight some practical complications in some of these strategies. State-of-the-art power optimization strategies can be categorized into heuristic control methods and static and dynamic optimization methods [6]. EV drivetrain energy efficiency is a function of EV system design, individual HV subsystems efficiency, and torque control strategies in different load demands. Therefore the torque control strategies alone are not sufficient. It is possible to command the optimal torque based on the battery characteristics and load demand to avoid unnecessary high discharge current, which accelerates depletion of the battery; but this is going to change the pedal feel for the driver, which is not suitable. Therefore OEM manufacturers give this option to the drivers to select whether they want to use eco-drive modes. Surveys show that eco-drive mode makes drivers be more looking at the traffic ahead to manage vehicle speed rather than driving with frequent unnecessary accelerations and brakes [7].

7.2.1.2 Brake mode torque mapping

There are two scenarios in brake mode based on the possibility to do regenerative braking or not. There is a growing interest in regenerative strategies because it is wrongly believed that regenerative braking can increase the mileage effectively. Based on practical experience and according to the collected data, authors would like to emphasize that in heavy-duty vehicles, the recovered energy through braking/deceleration is relatively low in comparison to the required energy to drive/accelerate the vehicle.

The optimal regenerative energy recovery depends on the initial and final braking velocity, motor efficiency, battery efficiency, and the motor maximum braking torque constraints [1, 2]. Among all the above constraints, the battery charging current limitation effectively reduces battery efficiency. The maximum charge current of the Li-ion cells used in heavy-duty applications is generally half of the maximum discharge

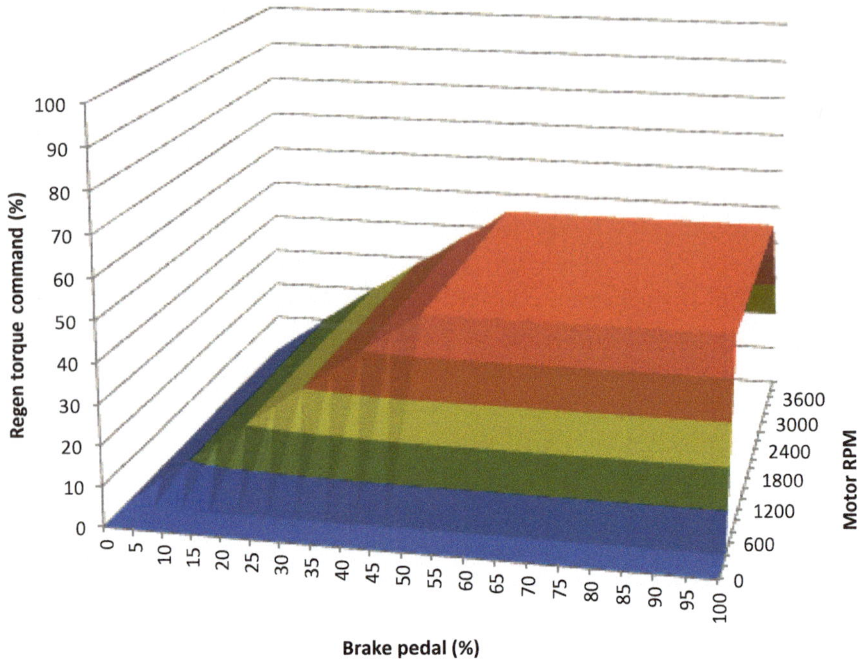

Fig. 3 A typical regenerative torque mapping graph based on brake pedal position and electric motor RPM.

rate. Therefore the maximum negative (regenerative) torque that can be commanded to the motor controller will be 50% of the maximum positive (motoring) torque. Besides, batteries cannot accept maximum charge current if their state of charge (SoC) is above the certain limit (usually 80%) and regenerative braking must be disabled for SoC above 95%.

Another major constraint is the short duration and low frequency of braking conditions in comparison to the acceleration demands. Motor regeneration efficiency is very low at low speeds; therefore usually regeneration will be active, while the vehicle speed is above a certain limit (to be calculated based on the motor RPM and gear ratio—usually 30 km/h). Fig. 3 shows a typical regenerative torque mapping graph based on brake pedal position and motor RPM. As shown in Fig. 3, the maximum regenerative torque is limited to 50%, and in low motor, RPM is zero.

7.2.1.3 Torque limiting modes

Certain safety factors limit the final torque command to the motor controller either in motoring or regenerative braking torque commands. The main factor is the heat generated in the battery, motor controller, and electric motor owing to high currents involved in acceleration or deceleration. Any torque limiting action must be informed as a warning to the driver (WtD).

Most HV subsystems must be cooled either by forced air or by circulating coolant. In general, there is a radiator and a cooling pump that circulates the cooled coolant (water/glycol) in the motor, inverter, DC/DC converter, and battery charger. There are air-cooled systems as well in low-power EV applications but not in heavy-duty applications. Liquid-cooled battery packs are not that common in heavy-duty EVs mainly owing to the size and cost involved. Suitable battery cells for heavy-duty applications are usually high-energy density type cells and have low discharge current rating. Connecting more battery cells in parallel is the solution to deliver the required current. Also, in comparison to passenger cars, there is more space available to locate batteries in buses or trucks, which helps to increase the energy of the battery and consequently vehicle mileage.

The torque limiting parameters to be considered in various EV subsystems are summarized in Table 1. The given limit values are routine checks in heavy-duty applications, however, they might change depending on the quality of components used, EV developer preference, and application requirements. Torque values are represented as a percentage of the maximum available torque.

7.3 Drivetrain motor controller parameterization

Advanced EV motor controllers are usually designed to drive various motor types such as AC induction motor (IM) and permanent magnet synchronous motors (PMSMs), such as IPM and SPM. These controllers are already programmed to control the electric motor based on the torque command on CAN; however, certain parameters need to be set in the program according to the selected motor, battery specifications, and performance requirements. Advanced EV motor controllers have up to 300 internal parameters that need to be set for optimal motor control in the below categories:

- Electric motor selection and operation limits.
- Electric motor encoder/resolver setup.
- Electric motor parameters.
- Current controller setup.
- Field weakening settings.

The motor inverter is a local controller that directly controls the electric motor and EV drivetrain. Optimal performance of the motor according to EV drivetrain torque mapping strategies is critical. Some of the

Table 1 Torque limiting values in heavy-duty electric vehicles.

Subsystem	Parameters	Actions
BMS	SoC < 10%	Limit the torque to 50% + WtD
	SoC < 5%	Limit the torque to 30% + WtD
	SoC < 2%	Limit the torque to 0% + WtD
	Cell temp. > 55°C Cell temp. < −5°C	Limit the torque to 50% + WtD
	Cell temp. > 60°C Cell temp. < −10 °C	Limit the torque to 30% + WtD
	Cell temp. > 65°C Cell temp. < −15°C	Limit the torque to 30% + WtD
Motor inverter DC/DC converter	IGBT temp. > 65°C PCB temp. > 65°C	Limit the torque to 50% + WtD
	IGBT temp. > 70°C PCB temp. > 70°C	Limit the torque to 30% + WtD
	IGBT temp. > 75°C PCB temp. > 75°C	Limit the torque to 0% + WtD
Electric motor	Temp. > 120°C	Limit the torque to 50% + WtD
	Temp. > 130°C	Limit the torque to 30% + WtD
	Temp. > 140°C	Limit the torque to 0% + WtD

parameter settings in the motor controller are similar to what has been discussed in torque mapping strategies and add redundancy to the overall EV driveline system.

7.3.1 Selecting the motor and setting the operation limits

Selecting the motor is the first step in setting up the motor controller. PMSMs and IMs are the most common types that have been used in heavy-duty EVs so far. Since the control processes of these motors are different, then some of the parameters that need to be set are different as well. For example, slip or magnetization current settings are only required for IMs only. Therefore IM and PMSM have been used to distinguish the dedicated parameters to each motor from standard parameters. The primary operation limits and cutbacks that need to be set, which results in limiting the power by the motor controller, are explained further:

- **Acceleration and deceleration limits**: These limits control the maximum possible rate of change of speed. In heavy-duty vehicle applications, the value is usually set between 1000 and 3000 RPM per second depending on the rated and maximum speed of the motor and gear ratio. The inverter maintains the maximum acceleration by limiting the commanded torque from the SC.
- **Maximum motoring and regenerative torque limits**: They are to be set according to the electric motor torque characteristics given by the manufacturer. As discussed earlier, maximum regenerative torque is usually between 30% and 50% of the maximum motoring torque.
- **Maximum motoring and regenerative power limit**: This is to be set according to the electric motor power characteristics given by the manufacturer and operation requirements. MC monitors the input DC voltage and current to calculate the power. Regenerative power must be limited to the maximum charging capability of the battery.
- **Maximum forward and reverse motor RPM**: This is to be set according to the electric motor speed characteristics given by the manufacturer and operation requirements. The reverse speed limit is usually set to be lower than maximum forward speed in heavy-duty vehicles. Some motor controllers have an option to set warning speed limits as well.
- **Motor temperature limits**: They are to be set according to the electric motor speed characteristics given by the manufacturer. The MC limits the torque when the temperature is above the defined limit and set the motor temperature warning flag that is communicated over CAN.
- **IGBT module and PCB temperature limits**: They are to be set according to the motor controller characteristics given by the manufacturer. The MC limits the torque as soon as the temperature is above the set limit and set the IGBT/PCB temperature warning flag that is communicated over CAN.
- **Minimum and maximum DC-link voltage limits**: They are to be set according to the battery pack maximum and minimum voltage limits. MC limit the torque to zero when the DC link voltage is not within the defined limits.
- **Maximum DC-link current limit**: This is to be set according to the operation requirements. MC limits the torque to zero when the DC link current is above the defined limit.

7.3.2 Setting up motor encoder/resolver

Internal motor speed sensor settings are one of the most critical tasks because it directly affects the control of the motor. In AC induction motors, the speed sensor is used for the motor speed and torque control. In permanent magnet motors, the speed sensor is not only used for speed and torque control purposes but also the commutation of the motor is controlled according to the rotor position that is detected by the internal sensor.

Incremental (ABZ) Encoders and Sine/Cosine Resolvers are the most common in EV electric motors. Some low-cost, low-power permanents magnet motors use Hall effect sensors that are not suitable for high performance applications. Incremental encoders output is two square waves that have 90 electrical degree phase difference. The direction of rotation is determined based on the sign of the phase angle between the square waves. Incremental encoders are commonly used in AC induction motors. Resolvers' output is two Sine and Cosine waves that have 90 electrical degree phase difference. Therefore the exact position of the rotor can be calculated based on the degree of rotation. The direction of rotation is determined based on the sign of the phase angle between the Sine and Cosine waves. Resolvers are commonly used in permanent magnet motors. The encoder gain and offset setup process in MC can be done using below steps:

- Choose the correct encode/resolver type and excitation source.
- Rotate the motor shaft in the forward direction around 5% of the maximum motor speed using a dynamometer and monitor the encoder signals. No HV connection to the motor is required.
- Set the encoder/resolver delay compensation to zero (if applicable).
- For incremental (ABZ) encoder:
  - Set the number of encoder lines.
  - Number of encoder lines is equal to the total number of A and B signal edges.
- For resolvers:
  - Set the offset values for Sine and Cosine waves to have symmetric signals with a 90-degree phase difference.
  - Set the gain values for Sine and Cosine waves to have the signals with an amplitude of 1.
- Rotate the motor shaft in the reverse direction and make sure the encoder/resolver waves are according to the specifications.
- Check the rotor angle value on CAN for rotation direction confirmation.
  - If the rotor angle increases/decreases, then the motor is rotating in forward/reverse direction, respectively.

The delay compensation is to adjust the phase lag of the encoder/resolver output signals at high speed. The process is same as explained to adjust the offset value at high speeds (70%–80% of the maximum motor speed). Therefore it is recommended to tune the speed and current controller parameters first before running the motor at high speed.

7.3.3 Setting motor parameters

Some basic parameters are common for both permanent magnet motors (IPM and SPM) and AC induction motors. However, each motor type has their dedicated parameters that need to be set too. These basic parameters will be used in speed and current control algorithms:

- **Pole pairs**: It defines the number of electric motor pole pairs; to be added from the electric motor datasheet given by the manufacturer.
- **Stator resistance**: It is the stator winding phase resistance plus the resistance of the HV cable from that phase to the motor. Phase resistance cannot be measured using a multimeter. A constant current has to be supplied through two phases of the stator winding to measure the voltage drop across two phases. The phase resistance is half of dividing the voltage to the current.
- **Rotor resistance (IM)**: It is the resistance of the AC induction motor rotor; to be added from the electric motor datasheet given by the manufacturer.
- **Stator and rotor leakage inductance (IM)**: It is to be added from the electric motor datasheet given by the manufacturer. If the values are not available in the datasheet, then the typical value could be between 0.15 and 0.25 mH.
- **Maximum motor phase current**: It is the maximum phase current (Arms) of the motor; to be added from the electric motor datasheet given by the manufacturer.
- **Torque constant**: It is the ratio that defines the relationship between the motor current and generated torque; to be added from the electric motor datasheet given by the manufacturer.
- **Back-EMF constant**: It is the ratio that defines the relationship between the motor speed and the induced back-EMF voltage to the windings; to be added from the electric motor datasheet given by the manufacturer. If the value is not available in the datasheet, then it can be calculated by running the motor at 10% of maximum speed using dynamometer, measuring the induced back-EMF voltage and then dividing it by speed in rad/s.

- **Minimum magnetization current (IM)**: The minimum current required to magnetize the induction motor rotor at zero torque request; to be added from the electric motor datasheet given by the manufacturer. Typically, it is 25% of the maximum motor phase current (Arms).

7.3.4 Current controller setup

The most commonly used control algorithm in the motor controller to control torque/current is field-oriented control (FOC). The closed-loop current control system is one of the main parts of FOC. The scope of this chapter is not to discuss motor control algorithms; we just review the required current control tuning parameters that need to be set up in MCs. The main objective is to set the current PID controller gains in such a way that the motor current (*Id* and *Iq*) reaches the reference targets quick and without overshoot. The current controller gains tuning is done in the lock rotor situation. The *Id* and *Iq* values are derived from the product of the Clarke and the Park transforms that simplify the analysis of three-phase circuits [8]:

- **Hysteresis current limits** defines the boundaries that controller maintains the actual motor current around the commanded reference current.
- **Maximum current difference warning limit** defines the difference limit between the actual current and the reference current. If the difference is more than the defined limit, a warning flag will be set and communicated on CAN.
- **Maximum *Id* and *Iq* limits** define the maximum limits for *d*-axis and *q*-axis current regulator.
- **Current sensor filter contact** is the time constant of the first-order filter that is applied to the current sensor signal. Typically, this delay could be from 1 to 5 ms depending on the current sensor type.
- ***Id* and *Iq* proportional gains (*Kp*)** define the proportional gains of *Id* and *Iq* axis current PID controllers.
- ***Id* and *Iq* integral gains (*Ki*)** define the integral gains of *Id* and *Iq* axis current PID controllers.
- ***Id* and *Iq* derivative gains (*Kd*)** define the derivative gains of *Id* and *Iq* axis current PID controllers. Some motor controllers have PI current controller only, and there is no *Kd* gain to be set.

- ***Ip* and *Iq* nonLinear *Kp* gain limit (IM)** defines a minimum limit the *Kp* value decreases linearly below that.
- ***Ip* and *Iq* nonlinear *Kp* gain slope (IM)** define the slope of decreasing *Kp* below the nonlinear *Kp* gain limit.

7.3.5 Field weakening settings

The induced back-EMF voltage in the stator windings goes higher than *d*-axis and *q*-axis voltages at high speeds; therefore *q*-axis current will be zero, and the motor generates no torque. In PMSMs, adding *d*-axis current reduces the induced back-EMF voltage; therefore, *q*-axis current flow and motor can generate torque. In IMs, the *d*-axis current produces magnetization current in rotor; therefore, decreasing the *d*-axis current reduces the induced back-EMF voltage. The process of controlling *d*-axis current to generate maximum possible torque at high speeds is called field weakening (FW):

- **Maximum *Ud* and *Uq* limits** define the maximum limits for induced back-EMF voltage on *d*-axis and *q*-axis that field weakening starts if the induced voltage is above these limits; to be added from the electric motor datasheet given by the manufacturer.
- **Minimum speed to start FW** defines the speed limit that field weakening starts above that speed; to be added from the electric motor datasheet given by the manufacturer.
- **FW *Id* change rate** defines the slope of change of *d*-axis current in the flux weakening state.
- **Maximum FW *Id* limit (PMSM)** is the maximum value of *d*-axis current in the flux weakening state.

7.4 Summary

Various EV drivetrain (electric motor and its inverter drive) operations mode and their respective torque control strategies have been explained in this chapter. Heavy-duty EV drivetrain torque mapping strategies in drive mode and brake mode have been explained, and zero torque conditions have been deliberated. Required parameter settings in the inverter drive for optimal performance of the motor have also been explained, and the methods to set them have been discussed.

References

[1] B. Li, Z. Yu, C. Jin, Z. Zhu, Torque allocation algorithm of distributed driving electric vehicle based on energy consumption optimization, in: 2015 7th International Conference on Intelligent Human-Machine Systems and Cybernetics, Hangzhou, 2015, pp. 319–323. 0.1109/IHMSC.2015.46.

[2] Y. Zhang, W. Yuan, R. Fu, C. Wang, Design of an energy-saving driving strategy for electric buses, IEEE Access 7 (2019) 157693–157706, https://doi.org/10.1109/ACCESS.2019.2950390.

[3] B. Li, A. Goodarzi, A. Khajepour, S.-K. Chen, B. Litkouhi, An optimal torque distribution control strategy for four-independent wheel drive electric vehicles, Veh. Syst. Dyn. 53 (8) (2015) 1172–1189.

[4] M. Dizqah, B. Lenzo, A. Sorniotti, P. Gruber, S. Fallah, J. De Smet, A fast and parametric torque distribution strategy for four-wheel-drive energy-efficient electric vehicles, IEEE Trans. Ind. Electron. 63 (7) (2016) 4367–4376.

[5] L. Li, C. Yang, Y. Zhang, L. Zhang, J. Song, Correctional DP-based energy management strategy of plug-in hybrid electric bus for city-bus route, IEEE Trans. Veh. Technol. 64 (7) (2015) 2792–2803.

[6] W. Wang, Z. Zhang, J. Shi, C. Lin, Y. Gao, Optimization of a dual-motor coupled powertrain energy management strategy for a battery electric bus based on dynamic programming method, IEEE Access 6 (2018) 32899–32909, https://doi.org/10.1109/ACCESS.2018.2847323.

[7] M. Günther, N. Rauh, J.F. Krems, Conducting a study to investigate eco-driving strategies with battery electric vehicles—a multiple method approach, Transp. Res. Procedia 25 (2017) 2242–2256.

[8] S.J. Chapman, Electric Machinery Fundamentals, fourth ed., McGraw Hill Publishing, 2010, ISBN: 0-07-246523-9.

CHAPTER 8

Battery Management System: Charge Balancing and Temperature Control

8.1 Introduction

Voltage and current throughput of a single lithium-ion (Li-ion) battery cell cannot meet the requirements of energy-intensive applications such as electric vehicles (EVs). Anode and cathode materials define the working potential range for each battery. State-of-the-art Li-ion battery cells cycle between 2.4 and 4.2V [1]. A power network is, therefore, created by electrically connecting many single cells in series and in parallel, which leads to a battery pack design. Noteworthy to mention here is that the performance of battery packs degrades at a faster rate in comparison to single cells because performance characteristics of different cells in the battery pack tend to diverge as they age [2].

Ideally, individual battery cells are expected to have identical characteristics and, therefore, should react equally to a stimulus. In practice although, cell-to-cell variations or inconsistencies are quite common. For example, some cells may operate at a higher voltage than others owing to differences in their temperatures, whereas in others, the observed voltage variations and dissimilar discharge rates could be owing to the differences in their internal resistances. Factors responsible for the cell inconsistencies can be both internal factors and external causes. Internal factors are introduced in the system because of inherent limitations of the cell manufacturing processes, material properties, fabrication, and assembly techniques. They include variations in physical volume, internal resistance and impedance, self-discharge rate, and capacity. On the other hand, external causes refer to issues such as the difference in charging and discharging rates of different cells owing to a preexisting temperature gradient in the battery pack and varying depths of discharge. Coupled effect of the external factors and the internal factors can exacerbate the cell-to-cell variations during regular battery pack operation [3–5].

Heavy-duty Electric Vehicles
https://doi.org/10.1016/B978-0-12-818126-3.00005-1

In a series connection, overall (dis)charge capacity is limited by the weakest cell in the link, that is, the cell with the lowest cut-off voltage(s). As a result, the capacity utilization rate of the battery pack is severely reduced. In case of considerable inconsistency, some of the battery cells can get overcharged during the charging process or deeply discharged during the discharging step. The operational reliability of Li-ion batteries can only be guaranteed when they are operated within a safe zone, that is, manufacturer-specified window. Overcharging and deep discharging are both considered a form of abuse for the present generation Li-ion batteries [6]. Abuse of Li-ion batteries can start an uncontrollable chain of chemical reactions that generate heat and gases like carbon dioxide (CO_2). The released heat makes the cell temperature rise. Increased temperatures accelerate battery aging and degradation phenomenon. In addition, gas/CO_2 generation causes pressure buildup inside the cells, which if not released on time can cause them to explode. For all these reasons, battery manufacturers carefully screen and sort the cells in groups before proceeding to battery pack assembly. Nevertheless, minor differences remain and get magnified gradually as the pack is used, causing safety issues [7]. Controlling internal factors beyond a certain engineering limit has severe financial implications. It is thus critical to monitor and regulate the external factors that influence the cell performance closely [8, 9]. Battery management system (BMS), also known as the battery monitoring system, is a tool to do just that.

8.2 Battery management system

The charge capacity of the cell at any specific temperature refers to the total energy that gets stored as the cell is charged (by a current that is per manufacturer's specification) from a lower cut-off voltage to the upper cut-off voltage and vice versa during discharge. It is a general understanding that battery topology acts as a voltage equalizer across parallel connections in the pack. It means that the total current in the pack is distributed among the parallel branches automatically in a ratio that ensures identical voltages across them. However, noteworthy to mention here is that the topology only balances the total voltage across parallel strings. It does not influence or equalize voltage of individual cells in one string where a string is defined as a series connection of cells. As a result, the upper and lower cut-off voltages of the weakest cell control the maximum charge/discharge capacities of a series connection [10]. Charge or discharge capacity of a battery pack can, therefore, be expressed as:

$$C_{\text{pack}} = \min\left(\text{SOC}_i \cdot C_i\right) + \min\left[\left(1 - \text{SOC}_j\right) \cdot C_j\right] \quad (1)$$

where C_i and C_j are the (dis)charge capacities of the *i*th and *j*th cells, respectively. Similarly, SOC_i, and SOC_j are state of charge (SOC) of the respective cells, and C_p is the (dis)charge capacity of the battery pack.

The first term in Eq. (1) describes the minimum remaining capacity of the weakest cell in a string, that is, series connection, while the second term denotes the minimum charging capacity of the weakest cell in that string. Suppose the same cell illustrates both the minimum remaining capacity and the minimum charging capacity in one string. In that case, the capacity of that cell defines the usable capacity of the battery pack [11]. Nevertheless, this is rarely observed in practice owing to embedded cell inconsistencies and nonuniform degradation behavior exhibited by them. Hence, the capacity of the battery pack is generally smaller than the capacity of any of the individual cells comprising the pack. The BMS monitors the external features affecting the battery pack during the use, identifies state parameters, and determines intercell variations over time. The balancing network is activated when a threshold is breached and initiates charge transfer to or from the cells that have either too low or too high capacities. Overall pack performance can be equalized after charge redistribution with various strings.

A BMS includes not only a set of electronic components but also a combination of functions and features that are necessary to meet the operational and safety requirements of the battery pack over the anticipated service life. It ensures that the battery's SOC remains in a prespecified window and monitors the change in its state of health (SOH). Protection against too high currents and large temperature variations is also a part of the BMS strategy. In short, BMS is critical for battery pack operation and provides the following functionality:

- measurement and control of voltage, current, and temperature
- SOC and SOH estimation
- fault detection
- data storage

This is illustrated in Fig. 1. BMS comprises the battery control unit (BCU), subcontrollers or cell monitoring units (CMU), an intelligent electric meter (IEM), and a data logger. A controller area network (CAN) bus is used as an internal communication unit. The IEM measures battery pack state

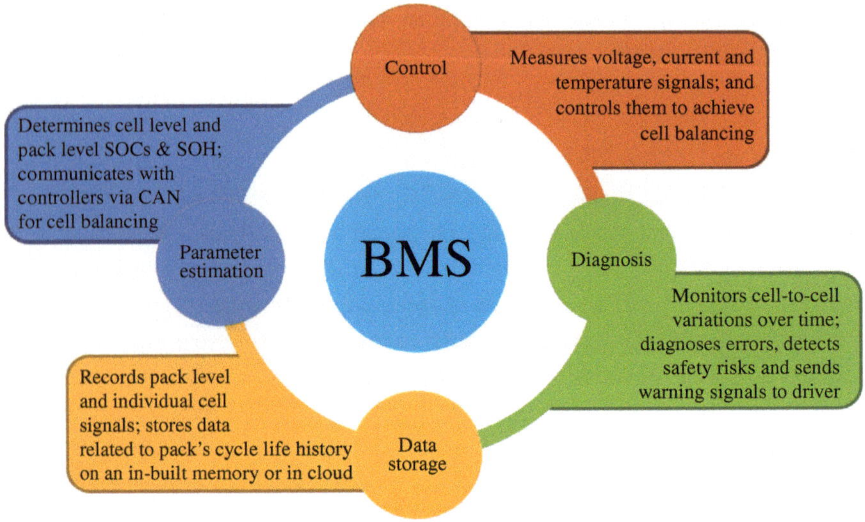

Fig. 1 BMS and its key functions.

parameters, that is, the total current and total voltage in the battery pack, and transfers the information to the CMU. Thereafter, the CMU performs a simple treatment of the temperature signals and the voltage signals of all the cells and sends the data to the CAN bus. The BCU receives the state signals from the CAN bus and responds by sending out control signals, which are then used to manage the whole battery pack.

BMSs could either have a centralized topology, distributed topology, or modular configuration. Centralized topology is a relatively low-cost system that uses one controller for managing all functions and is more common for small battery packs. However, the load controller used is cumbersome and has poor extendability. In a distributed topology, each individual cell has self-management capability because of an integrated control board. This topology enables measurements with higher precision, better antijamming, and reduces sensor harness complexity. However, hardware costs are substantial, and it gradually exacerbates cell inconsistencies owing to the power dependency of the controller on the corresponding cell. Last, the modular topology includes a master controller, multiple slave controllers, and an electric meter. The electric meter provides pack-level readings, that is, total voltage and total current in the battery pack with high precision. The slave controllers manage independent battery modules and measure voltages for each cell in those modules.

The master controller supervises these subsystems. It manages the data logger, electric meter, and various slave controllers through an internal CAN bus and communicates with the vehicle control unit (VCU) through an external CAN bus [12]. Modular topology provides good expandability and is preferred for EVs.

8.3 Charge equalization

In addition to the power network, interaction among individual battery cells happen through the communication network, the balancing network, and the safety network. The communication network is used for transferring information about each cell of the battery pack to the pack manager and to the user information console. The balancing network oversees charge equalization process and maintains a balanced SOC for the whole battery pack. Its topology can be different from that of the power network because only the components and connection that participate in the charge equalization process are considered part of the balancing network. Last, the safety network monitors fault detection tools and algorithms and activates pack reconfiguration panel/program as soon as a fault is identified. The hardware used in battery pack construction is network-specific. Consequently, the cost and complexity of the pack increase as the anticipated degree of functionality increases.

The balancing network has two distinct parts—hardware and software. Hardware mainly refers to the equalization circuit (EC), whereas software includes the equalization strategy (ES). The software is used for estimating the energy difference between various cells and the hardware enables the redistribution of the equivalent charge in the circuit. EC can be further classified as passive circuits and active circuits according to the power transfer mode and then according to the circuit topology as:

- cell bypass EC
- cell-to-cell EC
- cell-to-pack EC
- pack-to-cell EC
- cell-to-pack-to-cell EC

With regards to the ESs, they are designed as control optimization problems. They can be divided into three categories: (1) variable-based strategies; (2) objective-based strategies; and (3) algorithm-based strategies. The goal of each strategy type is to maximize the performance of the supported EC. The tree diagram presented in Fig. 2 illustrates this classification.

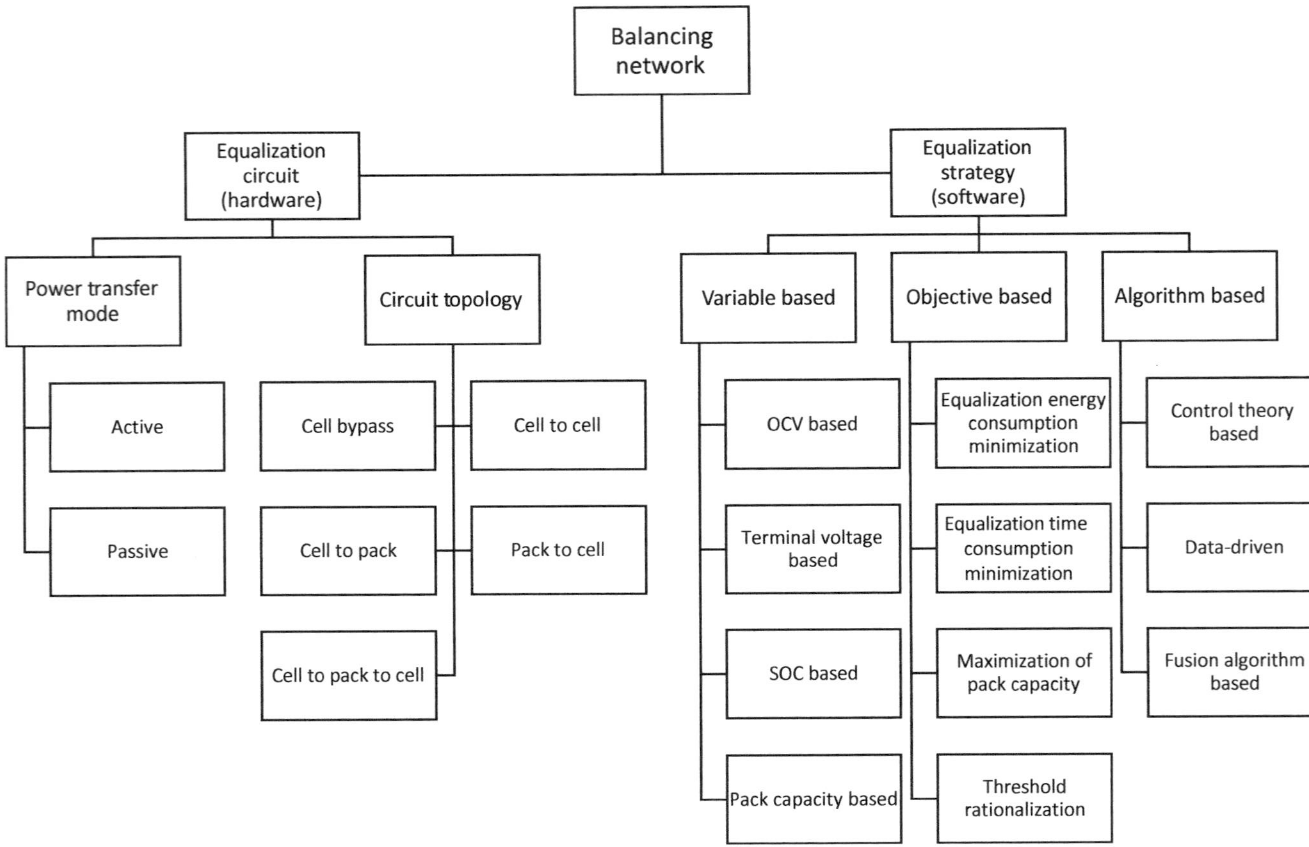

Fig. 2 Classification of balancing network.

8.3.1 Equalization strategy

The ESs comprise three parts: (a) mathematical model; (b) objective function; and (c) quality or performance indicators. For charge equalization systems, parameters such as open circuit voltage (OCV), terminal voltage, SOC, and capacity are accepted as inputs by the mathematical model along with other variables. However, there is always one output—equalization current. Design of the ES can revolve entirely around input variables. Such strategies are known as variable-based ES. Another approach is to base the design on optimization of the objective function, which can range from minimization of energy or time consumed in the equalization process to maximization of battery pack's usable capacity. Then, the most frequently used quality indexes for evaluating the performance of charge balance systems are transition time, accuracy, and stability. These are, in a way, dependent on control algorithms and their parameters. Such strategies that focus on optimization of quality indexes are called algorithm-based strategies.

Variable-based ESs can be further classified into voltage-based, SOC-based, or pack capacity-based strategies. Voltage-based strategy is the simplest of the three and applied widely because the terminal voltages can be measured directly. In contrast, implementation of both the SOC-based strategy and the pack capacity-based strategy is dependent on accurate and precise estimation of the SOC in real-time, which is challenging with the current state-of-the-art [13]. Hence their online application in EVs is relatively difficult.

The focus of the voltage-based ES is on achieving identical terminal voltages for each cell in a series string, within an acceptable tolerance range, of course. The balancing network makes use of a preset voltage threshold and compares it to the difference in terminal voltages of different cell groups in the string. Groups with a lower terminal voltage than the threshold value are then charged using an equalization current. Similarly, the equalization current functions as a discharging current for the cell groups operating at higher voltages than the threshold voltage. Threshold settings affect both equalization speed and stability. Thresholds can be either a single value threshold, multivalue thresholds, or adaptive thresholds. Single value threshold setting provides the shortest equalization time. However, caution must be used while defining a fixed value as a threshold. Single value thresholds can lead to an erroneous result and an increased energy consumption owing to repeated equalization attempts. Under such circumstances, equalization speed must be traded-off for a higher consistency of multiple thresholds and adaptive thresholds.

According to the battery state, threshold settings can be dynamically attuned to ensure reliable equalization in the shortest time possible.

Voltage measurement is a direct, quick, accurate, and reliable measurement, thereby making this a simple strategy from the implementation aspect. However, the computational load for executing voltage comparisons for each individual pair increases exponentially as the number of cells in the battery pack increases. Furthermore, several internal parameters such as internal resistance, capacity, and columbic efficiency, and external factors such as temperature, (dis)charge rate and SOH affect the cell voltage. Hence it is not an ideal strategy for balancing cells, particularly those of chemistry that exhibit voltage plateaus, for example, lithium iron phosphate (LFP) [14].

SOC represents the ratio of available capacity to the nominal capacity of the cell for specific (dis)charge conditions. As mentioned earlier, accurate knowledge of the SOC of the battery is essential for implementing SOC-based ESs. In comparison to the terminal voltage, OCV is regarded as a better marker of the internal state of the battery because dynamical electrochemical issues such as battery polarization and diffusion losses do not influence the OCV of the battery. Therefore a common practice involves balancing the cells by clipping voltage peaks and filling valleys in the voltage curves recorded during charging and discharging, respectively, based on the OCV-SOC relationship. Both the error probability and the computation requirements have been noted as insignificant for this approach. Moreover, it has been proposed to adopt an equalization current ratio-based equalization algorithm in the plateau region of the voltage curve and a voltage-based equalization algorithm outside the plateau area to overcome difficulties in balancing cells with a voltage plateau [15]. Regardless, there are some limitations, namely:

- The OCV-SOC curves may differ from cell-to-cell owing to intercellular variations introduced during the manufacturing process.
- The OCV measurement is a static process and works only while the cells are at rest—it is challenging to apply in real-time, that is, when the vehicle is in use, and the battery cells are being charged or discharged.
- Hysteresis can affect the OCV characteristics of a cell and as a result equalization process of the battery pack.

The OCV-based SOC estimation method was combined with ampere-hour (Ah) integral or the coulomb counting method [16, 17] to establish a relationship between dynamic SOC of the battery and the thermodynamic SOC by Feng et al. ESs for two separate conditions—equalization period and battery resting phase, were then developed to improve the overall effectiveness of the equalization process [18]. For a more accurate estimation of the cell's SOC, information about battery/environment temperature and aging behavior must be included in the analysis. Coupling battery thermal and battery aging models with the battery SOC estimation methods increases the complexity and computational costs for the process.

The balancing function works similarly to the voltage-based equalization technique. A preset SOC threshold value is provided to the BMS. If the measured SOC value for the cells exceeds this threshold, then the BMS activates a controlled discharge of these cells using an equalization current and vice versa for battery cells with the measured SOC values lower than the preset threshold. In the majority of the cases, the excess energy (from the cells operating at a SOC higher than the threshold) is transferred to the depleted cells to maximize the pack utilization rate. In others, it is dissipated as heat, which can be later used for regulating battery cell temperature in a cold environment.

Selecting SOC as the equalization variable offers some fundamental advantages. Some of them are listed below:

- The difference in the nominal capacities of individual cells becomes negligible—equalization current for each cell is in accordance to its SOC, and therefore all the cells attain either a fully discharged or a fully charged state simultaneously.
- Operation at different depths of discharge affects aging rates—because consistent SOC is maintained across the string, similar aging behavior can be expected for all the cells.
- Equalization time becomes short because it becomes possible to remove the energy and power difference among the cells in a single iteration.

Capacity-based equalization methods use discharge and charge capacities of the cells as equalization variables to establish charge balance among them. For passive equalization mode and active equalization mode, the theoretical maximum capacity of the battery pack is equal to the capacity of the weakest cell in the string and to the average of the capacities of all individual cells in the string, respectively. Active equalization is preferred because it circumvents the short plate effect of low-capacity cells in the string. Equalization current is estimated relative to the capacity of individual cells in a series string. Purpose of adopting pack capacity as strategy output is to ensure all the cells become fully charged or fully discharged at the same time. The BMS can then, for example, stop discharging the pack even before the lower cut-off voltage is reached for some cells and the available capacity or SOC becomes zero, which maximizes the battery pack capacity utilization. To alleviate the problems in estimating the total capacity of the battery pack accurately, Zheng et al. proposed using the charge-discharge voltage curves of a battery. Dissipative energy equalization based on total pack capacity was implemented, and capacity maximization was realized. Fuzzy logic control was applied to minimize the deviations of the equalized value from the theoretical capacity [19]. Nevertheless, the presence of voltage plateaus limits the accuracy of the technique. Therefore, Zhao et al. used a battery model, whose parameters were determined through nonlinear least squares method, for dynamic estimation of battery capacity [20]. Diao et al., on the other hand, concentrated on maximizing pack's releasable capacity by controlling the equalization current based on changes in battery's internal resistance, charging, and discharging rates/currents [21].

Using battery capacity as an equalization variable enables us to design the ES in line with operating conditions and maximize the functional capacity. Uniform current distribution and, as a result, uniform battery degradation across the battery pack, which translates to a prolonged cycle life for the battery pack, is a direct benefit of applying capacity-based ES. In addition to equalization time, energy consumed during the equalization process is less for the capacity-based ES in comparison to the voltage-based ES.

8.3.2 Equalization circuits

From the hardware perspective, ECs are classified based on energy conversion components used for equalization process. There are two major categories—passive ECs and active ECs. Passive ECs use energy dissipative mechanisms, for example, resistors connected in parallel to the battery cells. Resistors disperse excess energy stored in the cells operating above threshold as heat. In contrast, active ECs employ energy transferring devices such as capacitors, converters, inductors, and transformers in parallel for redistributing the excess charge equally among various cells. The former circuit topology is simple, economical, and reliable in service. However, it is a unidirectional circuit, that is, functions only during charging sequences and generates large amounts of heat even while using low equalization currents. As a result, passive ECs demonstrate long equalization time and low equalization efficiencies. In comparison, the latter EC topology is more advanced and allows maximum energy utilization [22, 23]. Reported efficiencies for all active ECs range between 85% and 97%. Speed, cost, complexity, etc. for different ECs are compared in Table 1.

Table 1 Comparison of different ECs.

EC	Current path	Speed	Control complexity	Circuit volume	Cost	Typical components in an EC of a battery pack with 96S1P configuration						
						Switches	R	C	L	T	Diodes	MWT
Switched capacitor	AC-2-C	Slow	Easy	Medium	Low	192	0	95	0	0	0	0
Single switched capacitor	DC-2-C	Slow	Medium	Small	Low	384	0	1	0	0	0	0
Double-tiered switched capacitor	AC-2-C	Medium	Easy	Medium	Medium	192	0	189	0	0	0	0
Cuk converter	AC-2-C	Medium	Complex	Medium	Medium	190	0	95	96	0	0	0
Buck-boost converter	AC-2-C	Medium	Complex	Medium	Medium	190	0	0	95	0	0	0
Flyback converter	DC-2-C	Slow	Complex	Medium	Medium	192	0	0	2	0	193	0
Single inductor	DC-2-C	Slow	Medium	Medium	Medium	194	0	0	1	0	1	0
Quasi resonant converter	AC-2-C	Medium	Complex	Large	High	190	0	95	190	0	0	0
Single transformer	C-2-P or P-2-C	Slow	Medium	Large	High	192	0	0	0	1	193	0
Multiwindings transformer	C-2-P or P-2-C	Slow	Complex	Large	High	96	0	0	0	0	1	1

The rightmost set of columns lists the number of various types of components that are needed to design a specific EC for a battery pack with 96S1P configuration. (R is resistor, C is capacitor, L is inductor, T is transformer, MWT is multiwindings transformer.)

Inductor or transformer-based controllers typically achieve equalization faster than the capacitor-based controllers, which in turn are faster than shunt-based controllers. However, careful sorting of cells preceding the pack design is crucial for all active circuit designs to derive energy gain that can justify the added cost and complexity of the circuit [24].

In one embodiment, known as adjacent cell-to-cell (AC-2-C), the active ECs work on shuttling principle, and the charge transfer happens continuously between adjacent cells at different states. The equalization takes places automatically without needing any voltage sensing and control system. Equalization speed for the AC-2-C circuit varies depending on the relative position of the unbalanced cell in the string and on the number of individual cells connected in that string. Another circuit type, known as directed cell-to-cell (DC-2-C), a basic control system coupled with voltage sensing, is used for arranging cells in descending order of their SOCs. Cells are targeted, and then energy is transferred starting from the cell with the highest SOCs to the next in line in a stepwise manner. The DC-2-C circuits balance a single cell at one time and, therefore total equalization times are quite long, especially for lengthy cell strings. This issue is resolved using a cell-to-pack (C-2-P) approach; wherein the voltage sensing system identifies the overcharged cells. Balancing procedure starts with the cell holding the maximum charge above a threshold. It transmits the extra charge to the entire pack simultaneously. Iterations required for balancing the whole pack using this method are lesser than those taken by the previous two schemes. Conversely, pack-to-cell (P-2-C) circuits disburse a small portion of pack's energy sequentially to all the undercharged cells. Cell with the lowest SOC is charged first in this sequence [25, 26].

In general, hardware circuit topology influences the equalization rate. Greater the charge imbalance or voltage difference, higher the equalization current. The current decays gradually and becomes zero when the energy balance is restored. Other factors that affect equalization time are EC topology, threshold setting, and control algorithm. For example, equalization time increases exponentially for a flying capacitor type EC as the equalization current decreases. Adopting too high equalization rates in order to minimize the equalization time can cause sharp spikes in equalization current and an undue increase in equalization energy consumption. Equalization energy consumption refers to a combined total of energies consumed by the battery pack and the EC. It can be contained to a minimum value by setting equalization energy consumption as the equalization objective.

Transformer-based ECs are quite common as well. A two-terminal transformer-based EC receives charge from an overcharged cell and transfers it to another individual cell with a lower than threshold SOC. In contrast, a multiple transformer-based EC absorbs energy from an overcharged cell (or a group of overcharged cells) and then distributes it to all cells of the battery pack. A critical benefit of transformer-based circuits is that electrical charge transfer happens through magnetic fields generated by transformer cores as opposed to electrical wires and cables used by other circuits. As a result, it becomes possible to electrically isolate individual cells of the battery pack from one another, which is useful for containing fault damage and for minimizing shock hazard. In addition, higher equalization rates can be achieved by activating all the transformers together. The number of intercellular connections is a limiting factor for the transformer-based charge balancing system though [27].

The number of intercellular connections, N_c, for connecting each cell of the pack with a total of N cells to every other cell in the pack can be represented as:

$$N_c = N(N-1)/2 \tag{2}$$

Complexities associated with the number of intercellular connections increases with the number of cells in the pack and can become unmanageable for large EV battery packs. A smaller number of connections is always preferable.

Balancing networks are usually static networks and may not share the same topology as the power network. However, balancing network topology significantly affects its ability to maintain uniform charge across all cells of the pack. Quinn and Hartley considered an example of a battery pack with a four-series, two-parallel configuration. Four different balancing network topologies, seen in Fig. 3, were compared to a case when no charge balancing is used.

The first topology included a network with series balancing on each parallel string of the battery pack. The second topology represented a ring arrangement where each cell is connected to its adjacent cells. The third topology included an entangled balancing network where the degree of connection was three, that is, each cell was connected to one other cell in the pack in addition to the two neighboring cells. The fourth and the final topology represented a network where each cell was connected to every other cell of the battery pack. Current capacities of all the five networks, that is, the total charge passing through the network per unit time when the voltage difference between the cells is constant, were kept identical by

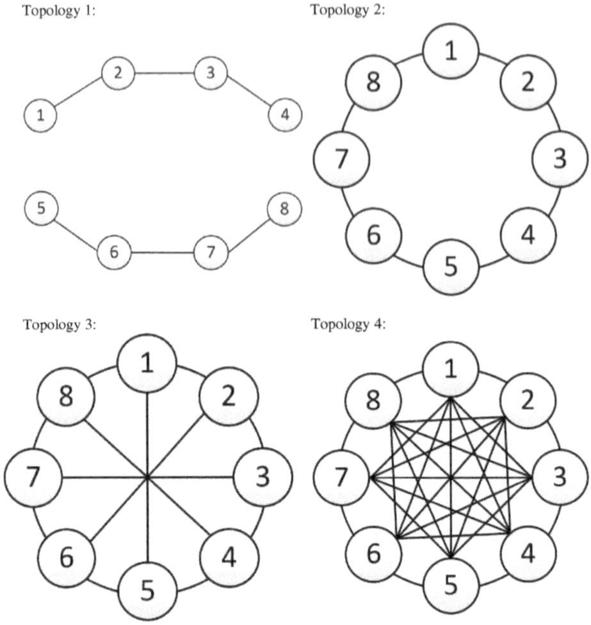

Fig. 3 EC topology variations—Topology 3 with a degree of connection 3 provides a better balance between performance and number of connections than other topologies.

scaling the transformer parameters in accordance with the total number of transformers comprising the network. Consequently, the average balancing current remains independent of the number of connections, and the observed performance variations can be attributed to the network topology. The number of steps involved in transferring charge from one cell to another in the first, second, third, and the fourth topologies was ∞, 4, 2, and 1, respectively. Similarly, spectral gaps or resiliencies of the four network topologies were 0, 0.586, 2, and 8, respectively.

Long-term voltage difference in the pack and the average exponential decay rate of this voltage difference varies with the variation in initial charge in the battery pack. Lower values of the voltage difference indicate better balancing over a lengthy period, and a higher value of the exponential decay rate suggests faster convergence to a stationary value. Given these two performance indicators, the first topology, that is, one with series balancing illustrated the worst performance, whereas the fourth topology or the network in which each individual cell is connected to every other cell in the pack was found superior to the other three network configurations. The third topology representing an entangled network with a degree of connection 3 provided an optimum balance between the performance delivered and the required number of connections [28].

8.4 Data storage

BMS in EVs and other large energy storage systems is required to store data relating to the cycle life history of their battery packs. The stored data can reveal information on usage patterns, performance, and capacity decay observed in the used battery packs. Analysis of the recorded data can support state estimation and fault diagnosis; off-line analysis of the historical data can also guide improvements in existing battery systems and design of more robust ones. Specified data are recorded through a data logger, a listen-only node of the CAN bus, on an in-built memory drive or on a secure digital memory card. Alternatively, the recorded data can be transmitted wirelessly to a cloud storage facility. However, the sampling frequency may need to be reduced either owing to limited on-board storage space or owing to poor wireless connectivity and slow data transmission speeds. Low sampling frequencies lead to data storage delay or long recording times, which in turn causes signal distortion and loss of information. In short, data quality is compromised.

Recorded data may include signals of the total current in the battery pack, total voltage across it, individual cell voltages, individual cell temperatures and environment temperature at specified locations and respective SOCs. The number of recorded signals increases from hundreds of signals at pack level to thousands when cell level details are also stored for battery packs of the size common in EVs. As an example, in a small battery pack with 96S1P configuration, approximately 400 signals are relating to total voltage, current, cell voltage, temperature, SOC, and other control signals are recorded. In comparison, the order of magnitude for the number of signals needed to be recorded in a battery pack with 240S20P configuration is 10,000.

Zheng et al. recorded 434 signals from an EV over 26 days. One-time recording of a single signal captured an average of 8 bytes memory space. A total of 7.4 Gigabytes (GB) were needed to store complete the recording of the 26 days period. Memory requirements for storing BMS data can be estimated using Eq. (3):

$$M = nTfm \tag{3}$$

where f is the sampling frequency, T is the total recording time, n is the number of recorded signals, and m is the average memory space required for storing an individual signal at one time, and M is the total disk space [12].

To an extent, battery pack design predetermines the number of signals that need to be recorded by the BMS. Dynamic usage patterns of EVs mandates a high sampling rate to facilitate accurate analysis, while a low sampling rate is essential for limiting the occupied disk storage space. A multifrequency recording principle was put forward by Zheng et al. to manage these contrasting requirements. The battery operates in three modes: (1) standby mode; (2) stationary charging mode; and (3) dynamic (dis)charging mode. Consequently, BMS signals have different dynamics. For example, current changes relatively little or not at all during the standby and the stationary charging modes and hence can be classified as a low frequency signal whilst the battery operates in these modes. However, it enters a high-frequency domain when the battery is being dynamically charged or discharged. Owing to the laws of physics, voltage and current belong to the same frequency domains and can be treated similarly. In contrast, it takes several seconds for the battery pack's SOC to change by 1% even during high-power (dis)charge use case [29]. Similarly, the high heat capacity of the battery pack and efficient thermal management systems (TMSs) minimize variation in battery properties owing to any thermal fluctuations. Therefore, both the SOC and temperature can be regarded as low-frequency signals. Multifrequency recording principle recommends adopting a high sampling rate for high-frequency signals and a low sampling rate for low-frequency signals. Different features of this approach are highlighted in Table 2.

Optimum recording frequency is the lowest frequency at which signal distortion is zero or negligible. To optimize the recording frequency, battery pack performance under all possible working scenarios is simulated, while the data logger operates at its maximum recording frequency. Discrete wavelet transformation (DWT) is used to reveal the most dynamic period in the recorded signal. The high amplitude frequencies are then distinguished from the recorded signal through the fast Fourier transformation (FFT) procedure. The threshold of the essential frequencies is determined. Thereafter, the optimal sampling frequency is identified through Shannon's sampling theorem, which specifies the minimum sampling rate for continuous-time signals [12]. It states that if all the frequencies making up a continuous-time signal $f(t)$ are smaller than ω Hz, then it can be reconstructed or completely recovered through uniform

Table 2 Features of multifrequency recording method.

Features	**Multifrequency recording**	
Signal type	Low frequency/static	High frequency
Working principle	Flexible recording frequency	Recording frequency optimization
Typical application	SOC, temperature, voltages and currents during standby and stationary charging modes	Voltages and currents during dynamic charging and discharging periods
Recording method	Signals and corresponding time recorded only when the difference between the incoming value and latest recorded value is greater than the threshold. Recording frequency is not preidentified but varies	The optimal sampling frequency is identified through FFT analysis to ensure no signal distortion or loss of information during the most dynamic phases

sampling at a rate $f_s \geq 2\omega$. Nyquist rate is the minimum sampling rate permissible under the sampling theorem, that is, $f_s = 2\omega$ [30, 31].

Now, accurate SOC estimation rely on the reliability of current and voltage measurements [32]. Therefore, the dependency of Ah and EKF-based SOC estimation methods on sampling frequency were analyzed. In this test, current and voltage sampling frequencies were reduced from 4 to 2 Hz. Between the two frequencies, the maximum current difference of approximately 0.04 Ah and a mean voltage difference smaller than 0.2 mV were recorded. It was observed that the information contained between the two sampling periods is mostly Gaussian white noise and does not affect SOC estimation accuracy.

A sampling frequency of 2 Hz can be used for recording dynamic current and voltage signals if available memory space is a real constraint. Zhou et al. proposed a frequency division model to tackle the challenge of limited storage space. They argued that battery pack's states change in a relatively short time frame, whereas cell inconsistencies become visible only over a long term. Consequently, time constants for measuring the change in battery pack states are of the order of milliseconds to seconds. On the other hand, state change at the cell level is a low-frequency phenomenon, and corresponding time constants range from hours to sometimes even days. Therefore the high-frequency phenomenon must be decoupled from low-frequency operations. Frequency division was applied through a mean-plus-difference model to achieve this objective. Herein, overall battery pack status, that is, average pack voltage and the total current were analyzed through a cell mean model (CMM). In contrast, a cell difference model (CDM) was used to investigate differences between individual cell voltages and internal resistances. Frequency of 0.4 Hz or lower was used for recording single cell data, and battery pack signals were retained as close to their original state as possible using a frequency of 2 Hz or higher. It was demonstrated that cell voltage signals recorded under the New European Driving Cycle (NEDC) at 0.2 Hz occupied only 28% of the original disk space with slight signal distortion. Root mean square error (RMSE) was 0.0097 V [33].

Network latency is another issue that affects the credibility of the signals recorded by the BMS. Latency refers to a delay in data transmission from the source to the receiver. Network latency can cause signals to become asynchronous and introduce an inherent time lag in the readings recorded by the data logger. As a result, recorded values will not indicate "true" state of the battery cell(s) and battery pack. This means the voltage difference measured between different cells at any specific time will not be accurate. Network latency can become increasingly problematic as the size of the battery packs and thereby the network increases. Cell inconsistencies drive the charge equalization process. However, signals reconstructed from asynchronous data will create artificial inconsistencies between different cells. Asynchronous currents and cell voltage measurements will lead to erroneous cell equalization, incorrect parameter and state estimation and difficulties in fault diagnostics. Hence synchronization is necessary for BMS data storage.

Recording a signal is a four-step process. The steps are—sampling, scheduling, transferring, and recording. Electric meter and slave controllers both contain analog-to-digital converters that transform the sampled signals to a digital form. Converted signals are then scheduled to be transferred over to the CAN bus. CAN messages are subsequently received by the data logger, wherein they are recorded as memories. Scheduling is done as per the priority of the message to be transmitted—pack-level messages get a priority over cell-level messages. Usually, one CAN message contains 72–128 bits, and its length can vary from 1 to

8 bytes. It takes 0.288–0.512 ms for broadcasting a single CAN message at a baud rate of 250 kbps. Furthermore, good "redundant" design practices require that CAN bus load rate is maintained around 30%. Therefore it would take approximately three times longer to transmit the whole message in practice.

In an EV BMS, an electric meter is responsible for recording the battery pack state parameters such as total voltage and current, as mentioned in Section 8.2. They are transferred in one CAN message. By virtue of this, total voltage and current are synchronous signals. In contrast, voltages across different cells of the battery packs are measured by different slave controllers, which then transmit the information in different CAN messages. Because different CAN messages are delivered one by one and on a priority basis, total voltage and current signals get recorded before the information related to the cell voltage. Consequently, the cell voltage signal and the current signal become asynchronized.

Kong et al. achieved online synchronization of cell voltage and current signals by using the original signal time t_o as the signal time instead of the time at which they were recorded t_r, which is typically used in majority of the models. Accordingly, upon receiving the CAN messages from the electric meter, data logger matches the total voltage and current signals to the most recently recorded time. As opposed to this, the signals received from the slave controllers are matched to a time when $j=0$, where j is a cycle counter initiated by the master controller. It runs from 0 to 20 in cases where a sending frequency of 20 Hz is used, that is, new command messages are sent by the master controller at an interval of 50 ms. Some latency may remain owing to a weak communication network or a wireless connection. This was addressed through frequency division equivalent circuit model-based parameter identification. A high-fidelity resistor-capacitor (RC) model with hysteresis was used in the high-frequency domain to represent the "mean" cell or the overall pack behavior, and simplified cell models were employed in the low-frequency region to represent differences between the individual cell and the "mean" cell. Proposed division simplifies the calculation and reduces the computational cost [34].

8.5 Thermal management

Significant progress has been made in the search for electrochemically stable battery materials. Nevertheless, the poor performance of Li-ion battery packs at elevated and low temperatures is still a major constrain. Exothermic electrochemical reactions characterize (dis)charging process in a battery cell [35]. Heat pockets can develop in battery packs owing to inadequate heat dissipation, especially while operating in hot climates or an insulating environment. Nonuniform temperature distribution alters (dis)charging characteristics of the battery packs and affects their cycle life [36, 37]. Managing the localized heat generation is critical for safe battery pack operation [38]. A TMS to minimize the effect of ambient temperature fluctuations on the battery performance is of vital importance.

Recommended guidelines are to regulate Li-ion battery cell temperature between 25°C and 40°C [39, 40]. In the recent years, several techniques have been developed for efficiently dissipating excess heat from battery packs. Broadly, they can be classified according to Fig. 4:

The medium used: Based on the working fluid used in cooling loop, a TMS can be classified as:

A. Air-cooled: unidirectional or reciprocating.

B. Liquid-cooled:

- **a.** Conventional liquid-cooled.
- **b.** Cold plate.
- **c.** Heat pipe.

C. Phase-change materials (PCMs).

D. Any combination of the above.

Power consumed: Active TMS includes a power-consuming equipment in the cooling loop, for example, evaporators, blowers, and pumps; otherwise, it is considered a passive TMS. In general, all active TMSs

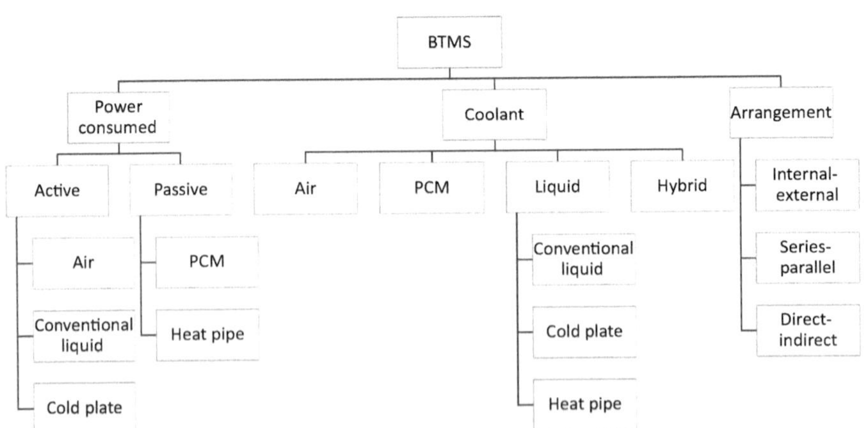

Fig. 4 Classification of different battery thermal management techniques.

transfer excess heat out of the battery packs in the form of sensible heat of the working fluid, while passive systems remove it as a latent heat.

Arrangement

A. Series-parallel.

B. Direct-indirect.

C. External-internal.

For commercial applications, there are two main approaches for designing a cooling system for battery packs: an air-based TMS or a liquid-based TMS. Basic construction and working principle of the two TMSs are depicted in Fig. 5. As a rule of thumb, the air-based TMS is more efficient for a small-size battery pack. In contrast, liquid-based TMSs are more effective in maintaining uniform temperature distribution in large battery packs operating at high discharge rates. Fins are a regular mechanical design feature used in an air-cooled system. In addition, reciprocating airflow can be applied to increase their efficiency. Liquid cooling systems are more complicated. Pumps are used for circulating coolant through a strategically designed plumbing in the battery pack, and a heat exchanger is included to reject the heat removed by the coolant to the external environment.

For example, a liquid coolant is passed through built-in mini channels of a metallic plate kept in close contact with the battery cells. The plate generally has high thermal conductivity and a flat shape, which makes its application easy in the case of battery packs made of the pouch and prismatic cells. However, the same shape makes it difficult to transfer heat from cylindrical cells using a cold plate.

Careful design and additional engineering are required to ensure a proper seal and minimize leakage of coolant in liquid-cooled TMSs. Fouling is another major issue that plagues liquid-based cooling systems. Consequently, all the components used should be corrosion-resistant. They should all be easily accessible to facilitate regular maintenance check-up and cleaning schedules. The system should also be able to withstand any volume and pressure changes owing to an increase in temperature of the coolant. A surge tank can be incorporated to relieve extra pressure created in the distribution manifolds owing to volumetric expansion of the coolant. It also supports purging of gases trapped in the system through continuous coolant circulation in the cooling circuit.

The heat transfer process can be divided into two stages—absorption and rejection. Absorption refers to heat transfer from battery cells to the coolant, whereas rejection describes the transfer of heat from the coolant to the

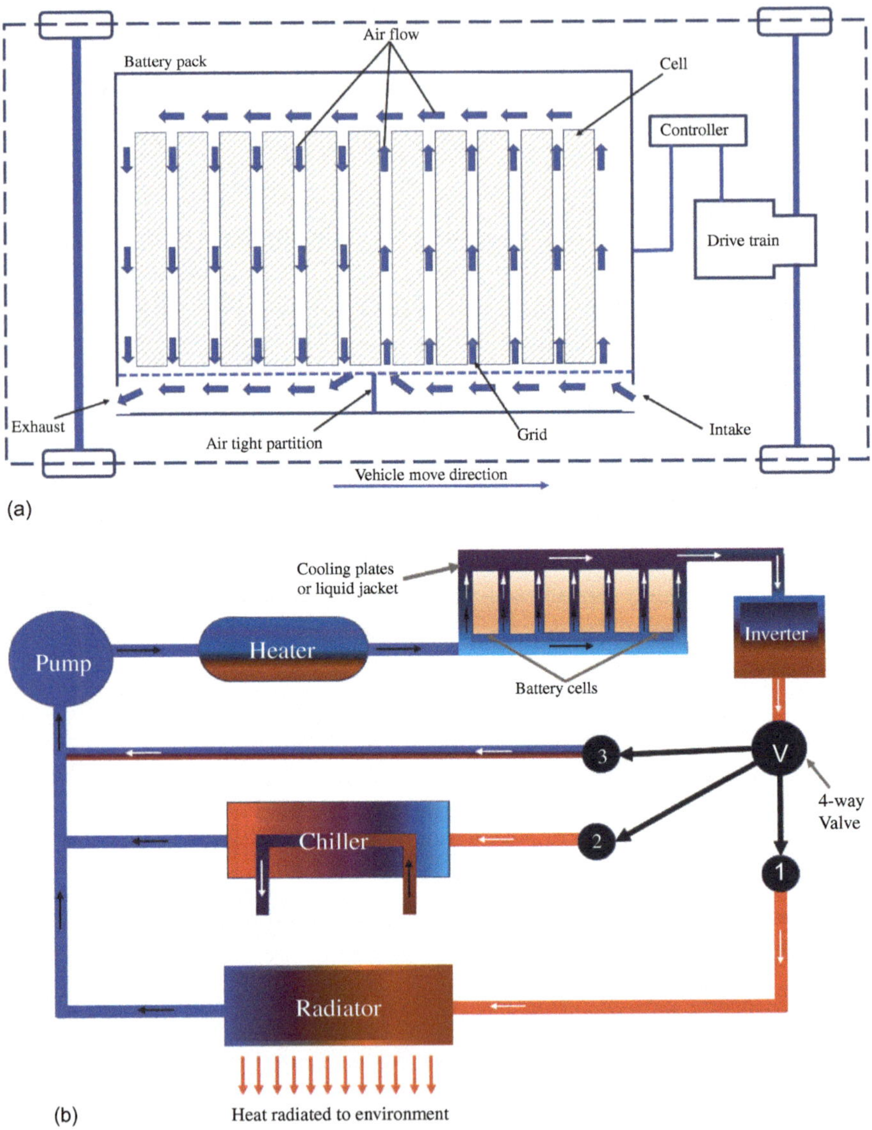

Fig. 5 Illustration of (a) an air-cooled thermal management system in an EV and (b) layout of a liquid-cooled thermal management system; *arrows* represent coolant stream (*black* = cold, *white* = heated coolant).

external environment. Absorption rate and efficiency are affected by the topology of the heat transferring surface, total contact area, heat transfer coefficient of the coolant, degree of fouling in cooling channels, and flow rate. Similarly, heat rejection efficiency is partly controlled by fin/heat exchanger geometry and design, flow rate and flow resistance, and ambient or environment temperature.

Inefficiency in heat transfer through liquid-cooled TMSs owing to incompatible geometries and marginalized contact areas can be removed through heat pipes or through PCMs. Heat pipes provide a compact cooling solution with high effective conductivity, while PCMs enable greater control over battery cell temperatures. However, the cost of the heat pipes and low thermal conductivity of PCMs needs to be considered in the selection of a suitable TMS for EV battery packs. In recent times, several other cooling techniques have been developed as well. These techniques offer many advantages, including significant energy and cost-saving potential, along with high scalability over traditional forced air- or liquid cooling methods. Table 3 provides a brief comparison of the traditional methods with these emerging techniques. For more details, the reader is referred to Ref. [41].

None of the TMSs compared in this table can alone fulfill the operational needs of a modular TMS. It can be inferred that a modular TMS would involve a combination of techniques. In other words, a modular TMS needs to be a hybrid system. The concept of thermal modularity is related to the concept of mechanical modularity. A modular TMS would ensure thermal independence of each battery cell and preserve its interchangeability. Interchangeable cells would promote mass production, which in turn would increase the cost competitiveness of EVs to diesel/petrol-powered vehicles.

There are several approaches for designing a TMS. The most appropriate approach depends on the desired level of sophistication, the availability of information, and the timeline/budget of the project. A step-by-step guide was presented by Pesaran [42, 43]. Basic steps are listed below:

1. Establish the desired battery pack thermal performance level, including
    - Average operating temperature.
    - Acceptable temperature variation between different cells.
    - Acceptable temperature variation in the pack.
    - Safety and packaging requirements for the vehicle.
2. Obtain module heat generation and heat capacity.
3. Perform a first-order steady-state energy and thermal analysis to estimate parasitic power required to operate the fan, pump, or heat exchanger.
4. Finalize the size of the TMS using a detailed transient analysis for the prediction of the battery module and pack behavior under different operating conditions.

Table 3 Comparison of traditional thermal management techniques such as air cooling and liquid cooling with emerging methods such as thermoacoustics-based thermal management system and magnetic refrigeration.

Criteria	Forced air	Liquid			PCM	Thermoelectric	Thermoacoustic	Magnetic
		Jacket	Cold plate	Heat pipe				
Ease of use	High	Low	Moderate	Moderate	High	Moderate	Moderate	Moderate
Integration	Simple	Difficult	Intermediate	Intermediate	Simple	Intermediate	Intermediate	Difficult
Energy efficiency	Low	High	Medium	High	High	Medium	Medium	High
Thermal gradient	High	Low	Moderate	Moderate	Low	Moderate	Moderate	Low
Cooling level	Small	Large	Medium	Large	Large	Medium	Medium	High
Regeneration rate	High	Medium	High	Medium	Low	High	Medium	High
COP @ room temperature	0.4–0.7	1.8–2.1	1.5–2.9	N/A	N/A	0.7–1.2	Up to 1.0	1.8
Maintenance	Low	High	Medium	Medium	Low	Medium	Low	Low
First cost	Low	High	High	High	Moderate	High	Low	Medium
Scalability	High	Low	Low	Low	High	Medium	Medium	High
Technical risks	Low	Medium	Medium	Medium	Low	Medium	High	Medium
Development state	Commercial	Prototype	Commercial	Prototype	Prototype	Commercial	Experimental	Experimental

5. Verification of the TMS design through experimental analysis confirming the battery performance and packaging requirements. The size of auxiliary components (fan/pump, HX, heater, etc.) is also determined. The significance of factors such as ease of operation, maintenance requirements, and reliability is evaluated. The control strategy for operating the TMS is also devised at this stage.

8.5.1 Heat generation estimation

One method to determine the amount of heat generated by battery cells is based on the principle of calorimetry. It allows to measure the quantity of heat exchanged between the system and its environment. Defining system boundaries is a prerequisite for using calorimetry. It is assumed that the system transfers all the generated heat to its environment. Owing to this assumption, the source of heating, in isolation, becomes immaterial to the problem. Temperature is measured at discrete times at one or more interior locations in the calorimeter to estimate the system's temperature history and thereby, heat generation.

Calorimeters are designed to function either under isothermal conditions or under adiabatic conditions. In an isothermal heat-conduction calorimeter (IHC), the battery surface temperature is maintained constant by keeping the battery surface in full contact with a large heat sink. However, IHCs tend to produce erroneous results at high discharge rates owing to the limited heat dissipation ability of the heat sink. As a result, their application is limited to coin cells and small cylindrical cells cycled at low discharge rates. In contrast, adiabatic calorimeters or accelerated rate calorimeters (ARCs) are used to evaluate battery heat generation rates in normal and in abusive environments. ARCs allow the battery cell temperature to increase over time while recording the thermal response of the calorimetric material because it transmits heat rejected by the battery cell during charging/discharging to a constant temperature heat sink. This information, coupled with the energy balance between the heat sink and the battery cell, is later used to assess the battery heat generation rates [44].

Estimation of heat generation through calorimeters is an example of inverse heat conduction problems (IHCPs). They are considered mathematically ill-posed and are sensitive to random instrumentation errors and the noise present in the experiment. A simple mathematical formulation of an IHCP is centered around Fourier's law of heat transfer and the fundamental law of conservation of energy. In this formulation, the problem is reduced to a single spatial dimension by considering lateral conduction in the calorimeter negligible in comparison to heat flow in the direction normal to the battery surface. The temperature distribution, T, as a function of time (t) in a slab of homogenous and isotropic calorimetric material is described by

$$\frac{\partial}{\partial x}\left(k\cdot\frac{\partial T}{\partial x}\right)=\rho\cdot c\cdot\frac{\partial T}{\partial t} \tag{4}$$

where k is the thermal conductivity of the calorimetric material, while ρ and c represent its density and specific heat, respectively. The initial condition of thermal equilibrium between the calorimeter and the environment is

$$T(x,0)=T_0(x) \tag{5}$$

and the boundary condition of no heat exchange with surroundings at $x=L$, where L represents the thickness of the calorimeter wall/slab is defined as

$$\frac{\partial T}{\partial x}=0 \text{ at } x=L \tag{6}$$

Temperature measurements at $x = x_1$, that is, the location of the sensor corresponding to discrete time steps, t_i, during the test cycle are given by

$$T(x_1,t_1)=Y_i \tag{7}$$

Interestingly, a one-dimensional IHCP with a single unknown surface heat flux at $x=0$, a known boundary condition at $x=L$ and a temperature history for one internal location can be converted to a case involving two separate problems. Now, for one of these problems concerning the portion of the slab spanning from $x = x_1$ to $x=L$ (say, body 2), the boundary conditions at both of its ends are known. As a result, heat transfer through this region of the slab can be analyzed as a direct problem. Heat flux entering body 2, q, which is practically equal to the flux leaving body 1 (portion of the slab from $x=0$ to $x = x_1$) through the surface $x = x_1$ is therefore calculated directly by solving the following equation:

$$q_{x_1}(t)=-k\frac{\partial T}{\partial x} \tag{8}$$

In addition, the battery cell can be assumed to be emitting heat equally from both faces. The total heat flux generated by it is therefore twice the calculated value [45].

Another method to approximate heat generation in batteries is based on an empirical function proposed by Tiedemann and Newman to describe irreversible polarization between the two electrodes of a battery cell:

$$\frac{I_t}{A} = Y(U - \phi_+ - \phi_-) \tag{9}$$

where $I_t/A = 0$ is the current density transferred locally from negative to positive electrode through a separator, U is the open-circuit voltage (OCV) of the battery, Y is the slope of the voltage-current ($V-I$) curve of the battery, and ϕ_+ and ϕ_- are potentials of positive and negative electrodes, respectively; their difference $(\phi_+ - \phi_-)$ is the measured voltage of the battery [46]. Polarization refers to the mechanism of displacing battery potential from its equilibrium during a chemical reaction. It can be divided into three categories: Ohmic polarization, concentration polarization, and activation polarization. Together they result in reduced cell performance.

The dependency of U on DOD can be expressed in a polynomial form as shown in Eq. (10). Third-order or cubic polynomials are generally preferred owing to their ability to capture important details and superior computational efficiency [47].

$$\text{OCV}_t = \sum_{m=0}^{3} A_m \cdot \text{DOD}^m \tag{10}$$

In the Tiedemann and Newman formulation, U represents voltage corresponding to a static condition when there is no current flow through the cell. It is mathematically calculated by the intercept of a tangent line on the polarization curve at $I_t/A = 0$. However, it is illustrated in Ref. [48] that a more accurate estimate can be made by replacing OCV with the equilibrium voltage. A battery is in an equilibrium state when the OCV change rate becomes less than 0.1 mV/30 min. It has been demonstrated that a commercial LFP battery cell operating in an ambient temperature of 27°C must remain in an open circuit state for at least 1 h in order to fulfill this criterion.

Furthermore, under a condition that the current used for both the charge and the discharge processes is the same, the irreversible heat for both these processes is assumed equal. Hence the reversible heat can be estimated from discharge and charge calorimetric data as follows:

$$Q_{\text{reversible}} = \frac{Q_{\text{discharge}} - Q_{\text{charge}}}{2} \tag{11}$$

Please note that battery modules and packs contain several battery cells combined in series and parallel. The number of cells in series and in parallel affects the electrical characteristics of the module. Therefore Eq. (9) or the battery polarization model cannot be applied directly to battery modules or packs. Equivalent values are first needed to be determined for each of the required parameters, for example, equivalent conductance for the module, overall voltage, and current, from basic electrical principles, which can then be substituted in the polarization model to estimate heat generation for battery modules.

8.6 Summary

Throughput of the state-of-the-art individual Li-ion cells is not enough to support an "electric powertrain." Therefore battery packs are needed. However, intercellular variations exist and get amplified because of the internal and external factors such as impedance difference, different charging and discharge rates, and temperature gradient. This has a negative effect on normal battery pack operation. A BMS is used to monitor these irregularities and control battery cell and pack performance in a close range.

This chapter provides a comprehensive discussion on key functionalities of BMS, which include charge balance, temperature regulation, and data storage. BMS architecture is described. The charge equalization process is elaborated in the context of various equalization strategies and circuits. Ability to record battery signals and store cycle life history data is a crucial requirement for BMS. However, a large number of signals that is typical of a battery pack from heavy-duty EVs and a limited on-board memory compromises this ability. To that end, the multifrequency recording method is introduced as a suitable data storage method.

In the end, temperature regulation methods suitable for modular battery packs are discussed. Estimation of heat generation from battery cells is necessary for sizing the TMS correctly. Basics of estimating heat generated by batteries through the application of calorimetry principles are explained. Another technique of approximating heat generation from battery polarization equation is also presented.

References

[1] F. Feng, X. Hu, J. Liu, X. Lin, B. Liu, A review of equalization strategies for series battery packs: variables, objectives, and algorithms, Renew. Sust. Energ. Rev. 116 (2019) 109464.

[2] Y. Zheng, M. Ouyang, L. Lu, J. Li, Understanding aging mechanisms in lithium-ion battery packs: from cell capacity loss to pack capacity evolution, J. Power Sources 278 (2015) 287–295.

[3] J. Kasnatscheew, M. Börner, B. Streipert, P. Meister, R. Wagner, I. Cekic Laskovic, M. Winter, Lithium ion battery cells under abusive discharge conditions: electrode potential development and interactions between positive and negative electrode, J. Power Sources 362 (2017) 278–282.

[4] C. Zhang, Y. Jiang, J. Jiang, G. Cheng, W. Diao, W. Zhang, Study on battery pack consistency evolutions and equilibrium diagnosis for serial-connected lithium-ion batteries, Appl. Energy 207 (2017) 510–519.

[5] A. Kulkarni, A. Kapoor, S. Arora, Battery packaging and system design for an electric vehicle, SAE Technical Paper 2015-01-0063, SAE International, Melbourne, Australia, 2015, https://doi.org/10.4271/2015-01-0063.

[6] S. Arora, A. Kapoor, W. Shen, A novel thermal management system for improving discharge/charge performance of Li-ion battery packs under abuse, J. Power Sources 378 (2018) 759–775.

[7] X. Feng, M. Ouyang, X. Liu, L. Lu, Y. Xia, X. He, Thermal runaway mechanism of lithium ion battery for electric vehicles: a review, Energy Storage Mater. 10 (2018) 246–267.

[8] S. Arora, A. Kapoor, W. Shen, Application of robust design methodology to battery packs for electric vehicles: identification of critical technical requirements for modular architecture, Batteries 4 (3) (2018) 30.

[9] S. Arora, W. Shen, A. Kapoor, Designing a robust battery pack for electric vehicles using a modified parameter diagram, SAE technical paper 2015-01-0041, SAE International, Melbourne, Australia, 2015, https://doi.org/10.4271/2015-01-0041.

[10] J. Meng, M. Ricco, A.B. Acharya, G. Luo, M. Swierczynski, D.-I. Stroe, R. Teodorescu, Low-complexity online estimation for Lifepo4 battery state of charge in electric vehicles, J. Power Sources 395 (2018) 280–288.

[11] L. Song, T. Liang, L. Lu, M. Ouyang, Lithium-ion battery pack equalization based on charging voltage curves, Int. J. Electr. Power Energy Syst. 115 (2020) 105516.

[12] Y. Zheng, M. Ouyang, X. Li, L. Lu, J. Li, L. Zhou, Z. Zhang, Recording frequency optimization for massive battery data storage in battery management systems, Appl. Energy 183 (2016) 380–389.

[13] P. Shrivastava, T.K. Soon, M.Y.I.B. Idris, S. Mekhilef, Overview of model-based online state-of-charge estimation using Kalman filter family for lithium-ion batteries, Renew. Sust. Energ. Rev. 113 (2019) 109233.

[14] S. Arora, Design of a Modular Battery Pack for Electric Vehicles (Doctoral thesis), Swinburne University of Technology, Melbourne, Australia, 2017.

[15] S. Li, S. Pischinger, C. He, L. Liang, M. Stapelbroek, A comparative study of model-based capacity estimation algorithms in dual estimation frameworks for lithium-ion batteries under an accelerated aging test, Appl. Energy 212 (2018) 1522–1536.

[16] K.S. Ng, C.-S. Moo, Y.-P. Chen, Y.-C. Hsieh, Enhanced coulomb counting method for estimating state-of-charge and state-of-health of lithium-ion batteries, Appl. Energy 86 (9) (2009) 1506–1511.

[17] J. Zhang, J. Lee, A review on prognostics and health monitoring of Li-ion battery, J. Power Sources 196 (15) (2011) 6007–6014.

[18] F. Feng, K. Song, R. Lu, G. Wei, C. Zhu, Equalization control strategy and Soc estimation for Lifepo4 battery pack, Diangong Jishu Xuebao/Trans. China Electrotech. Soc. 30 (1) (2015) 22–29.
[19] Y. Zheng, M. Ouyang, L. Lu, J. Li, X. Han, L. Xu, On-line equalization for lithium-ion battery packs based on charging cell voltages: part 2. Fuzzy logic equalization, J. Power Sources 247 (2014) 460–466.
[20] Y.F. Zhao, C.Q. Du, F.W. Yan, Energy equalization control scheme of power battery pack, Dianji yu Kongzhi Xuebao/Electr. Mach. Control 17 (10) (2013) 109–114.
[21] W. Diao, N. Xue, V. Bhattacharjee, J. Jiang, O. Karabasoglu, M. Pecht, Active battery cell equalization based on residual available energy maximization, Appl. Energy 210 (2018) 690–698.
[22] Y. Chen, X. Liu, H.K. Fathy, J. Zou, S. Yang, A graph-theoretic framework for analyzing the speeds and efficiencies of battery pack equalization circuits, Int. J. Electr. Power Energy Syst. 98 (2018) 85–99.
[23] N. Bouchhima, M. Schnierle, S. Schulte, K.P. Birke, Active model-based balancing strategy for self-reconfigurable batteries, J. Power Sources 322 (2016) 129–137.
[24] M.M. Hoque, M.A. Hannan, A. Mohamed, A. Ayob, Battery charge equalization controller in electric vehicle applications: a review, Renew. Sust. Energ. Rev. 75 (2017) 1363–1385.
[25] Y. Zheng, M. Ouyang, L. Lu, J. Li, X. Han, L. Xu, On-line equalization for lithium-ion battery packs based on charging cell voltages: part 1. Equalization based on remaining charging capacity estimation, J. Power Sources 247 (2014) 676–686.
[26] F. Baronti, R. Roncella, R. Saletti, Performance comparison of active balancing techniques for lithium-ion batteries, J. Power Sources 267 (2014) 603–609.
[27] J. Carter, Z. Fan, J. Cao, Cell equalisation circuits: a review, J. Power Sources 448 (2020) 227489.
[28] D.D. Quinn, T.T. Hartley, Design of novel charge balancing networks in battery packs, J. Power Sources 240 (2013) 26–32.
[29] C. Liu, Y. Wang, Z. Chen, Degradation model and cycle life prediction for lithium-ion battery used in hybrid energy storage system, Energy 166 (2019) 796–806.
[30] R.E. Blahut, Information theory and coding, in: W.M. Middleton, M.E. Van Valkenburg (Eds.), Reference Data for Engineers, ninth ed., Newnes, Woburn, 2002, pp. 1–31 (Chapter 25).
[31] E. Lai, Converting analog to digital signals and vice versa, in: E. Lai (Ed.), Practical Digital Signal Processing, Newnes, Oxford, 2003, pp. 14–49 (Chapter 2).
[32] X. Hu, F. Feng, K. Liu, L. Zhang, J. Xie, B. Liu, State estimation for advanced battery management: key challenges and future trends, Renew. Sust. Energ. Rev. 114 (2019) 109334.
[33] L. Zhou, L. He, Y. Zheng, X. Lai, M. Ouyang, L. Lu, Massive battery pack data compression and reconstruction using a frequency division model in battery management systems, J. Energy Storage 28 (2020) 101252.
[34] X. Kong, Y. Zheng, M. Ouyang, X. Li, L. Lu, J. Li, Z. Zhang, Signal synchronization for massive data storage in modular battery management system with controller area network, Appl. Energy 197 (2017) 52–62.
[35] R. Spotnitz, J. Franklin, Abuse behavior of high-power, lithium-ion cells, J. Power Sources 113 (1) (2003) 81–100.
[36] C. Lin, S. Xu, G. Chang, J. Liu, Experiment and simulation of a Lifepo4 battery pack with a passive thermal management system using composite phase change material and graphite sheets, J. Power Sources 275 (2015) 742–749.
[37] C. Zhu, X. Li, L. Song, L. Xiang, Development of a theoretically based thermal model for lithium ion battery pack, J. Power Sources 223 (2013) 155–164.

[38] D.D. MacNeil, D. Larcher, J.R. Dahn, Comparison of the reactivity of various carbon electrode materials with electrolyte at elevated temperature, J. Electrochem. Soc. 146 (10) (1999) 3596–3602.
[39] I.S. Suh, H. Cho, M. Lee, Feasibility study on thermoelectric device to energy storage system of an electric vehicle, Energy 76 (2014) 436–444.
[40] H. Kim, S.-G. Park, B. Jung, J. Hwang, W. Kim, New device architecture of a thermoelectric energy conversion for recovering low-quality heat, Appl. Phys. A Mater. Sci. Process. 114 (4) (2014) 1201–1208.
[41] S. Arora, Selection of thermal management system for modular battery packs of electric vehicles: a review of existing and emerging technologies, J. Power Sources 400 (2018) 621–640.
[42] A.A. Pesaran, S. Burch, M. Keyser, Institut Mech Engineers, An approach for designing thermal management systems for electric and hybrid vehicle battery packs, in: Vtms 4: Vehicle Thermal Management Systems, 1999, pp. 331–346.
[43] A.A. Pesaran, Battery thermal management in Ev and Hevs: issues and solutions, Battery Man 43 (5) (2001) 34–49.
[44] S. Arora, W. Shen, A. Kapoor, Neural network based computational model for estimation of heat generation in Lifepo4 pouch cells of different nominal capacities, Comput. Chem. Eng. 101 (2017) 81–94.
[45] S. Arora, A. Kapoor, Experimental study of heat generation rate during discharge of Lifepo4 pouch cells of different nominal capacities and thickness, Batteries 5 (4) (2019) 70.
[46] S. Arora, A novel technique for estimation of the solid electrolyte interphase film resistance for Li-ion batteries, in: Proceedings of the ASME 2018 International Mechanical Engineering Congress and Exposition, vol. 6A: Energy, Pittsburgh, Pennsylvania, USA, November 9–15, 2018, V06AT08A027, ASME, 2018, https://doi.org/10.1115/IMECE2018-87311.
[47] S. Arora, K. Tammi, A hybrid thermal management system with negative parasitic losses for electric vehicle battery packs, in: Proceedings of the ASME 2018 International Mechanical Engineering Congress and Exposition, vol. 6A: Energy, Pittsburgh, Pennsylvania, USA, November 9–15, 2018, V06AT08A025, ASME, 2018, https://doi.org/10.1115/IMECE2018-86111.
[48] S. Arora, W. Shen, A. Kapoor, Critical analysis of open circuit voltage and its effect on estimation of irreversible heat for Li-ion pouch cells, J. Power Sources 350 (Suppl. C) (2017) 117–126.

CHAPTER 9

Supervisory Control Systems for Heavy-duty Electric Vehicles

9.1 Introduction

Supervisory controller (SC) in electric vehicles (EVs) and hybrid EVs refers to the vehicle master controller that not only supervises the performance of all high-voltage (HV) subsystems in the vehicle but also controls the power flow from the HV battery to the electric motor. SC determines the actual output power according to the driver inputs and the vehicle operating condition [1].

The supervisory control system must manage the vehicle performance and the driveline components, including EV drivetrain. In this book, EV driveline is defined as the overall power transmission/conversion steps from the high voltage battery pack to the wheels that include all subsystems that are required for operating the vehicle. In contrast, the EV drivetrain is considered as the propulsion system that includes the electric motor and associated gearbox and/or differentials. Typical EV driveline subsystems included in a heavy-duty electric vehicle is as follows:

- **HV energy storage system** supplies the required energy to drive the vehicle. This is typically made up of multiple Li-ion cells packaged into suitably sized modules, which then are assembled into a single or multiple battery power packs, depending on the vehicle HV and LV electrical system design and mileage requirements.
- **Battery management system (BMS)** to protect Li-ion cells against overcharging, excessive discharge, and high and low operating temperatures. The BMS is a unit that monitors every individual cell's voltage, temperature, and the battery pack current. BMS uses this information to calculate battery pack state of charge (SoC), state of health (SoH), and Max. and Min. cell voltage/temperature values to be communicated to SC.
- **HV power distribution unit (HV-PDU)** distributes the power to the HV subsystems. This unit includes HV contactors and all the required

Heavy-duty Electric Vehicles
https://doi.org/10.1016/B978-0-12-818126-3.00009-9

protection and measurements, such as voltage, current, temperature, and isolation measurement. Depending on the vehicle control system topology, the HV-PDU can be controlled by the BMS or directly by the SC.

- **Motor controllers** drives the electric motor according to the commanded torque and speed from SC. Most available motor controllers in the market have control algorithms for various electric motor types, such as IPM, SPM, and ACIM, and can be parameterized by the user. Motor controller's software includes safety features to avoid potential electric faults in the motor.
- **DC/DC converter** to charge the 24-V low-voltage battery systems from the HV battery pack.
- **Electric air compressor** to provide the required air pressure mainly for brake applications and door operation in city buses.
- **Electric-driven power steering** to provide a required steering power to the driver. It is the same hydraulic steering system that is driven by the HV electric motor rather than a diesel engine.
- **Liquid-cooling system** for the HV subsystems includes a pump to circulate the coolant and a radiator with fans to cool down the coolant temperature. Most HV subsystems, including electric motors, motor controllers, DC/DC converters, require liquid cooling to optimize their performance.

9.2 Electric vehicle control system architecture

The vehicle control system has two-level control layers in general [1, 2]; low-level control that includes local controllers with required traditional feedback control [3], and the other high-level controller that is responsible for maintaining the optimal and safe operation of the vehicle and low-level controllers. Inverter control algorithms inside the motor controller to drive the motor, control algorithm inside DC/DC algorithm, and BMS are examples of the local controllers. Local controllers can be parameterized to achieve different objectives according to the required performance and high-level controller commands [4]. The communications between low-level and high-level controllers are mostly through controller area network (CAN) protocol; however, physical control signals such as pulse width modulation (PWM) and switched On/Off are also used in some cases.

The supervisory control solution (high-level controller) includes system functional control and vehicle operation modes algorithms and strategies. In addition to the above embedded control layers, SC includes diagnostics implementation to ensure safe vehicle operation [5]. These control layers are interconnected through complex algorithms, and a combination of all these strategies results in having a safe EV system that can achieve predefined operation requirements. The overall vehicle control system architecture is shown in Fig. 1.

As shown in Fig. 1, the EV supervisory controller includes three main control layers, namely

- System functional control.
- Vehicle operation modes control.
- Diagnostics and fault-handling strategies.

In an optimized EV design solution, the SC acts as the brain (central vehicle control unit) of the vehicle and implements all required strategies and algorithms associated with the above control layers. These embedded control layers are linked together accordingly to ensure the correct and safe performance of the vehicle.

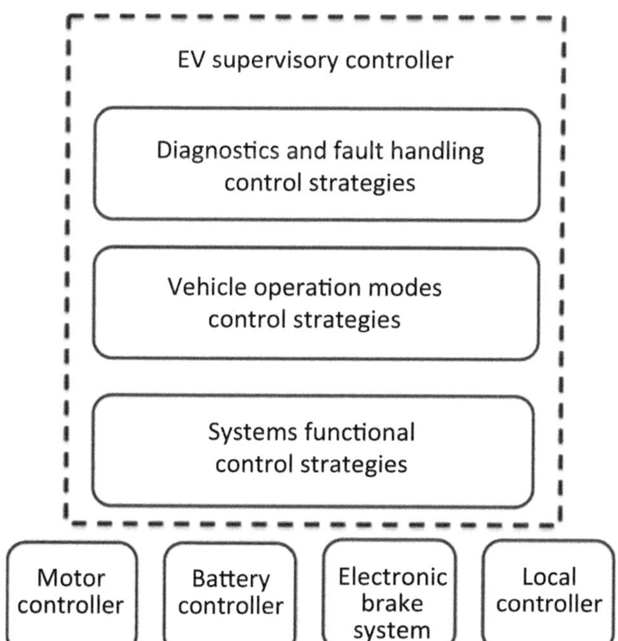

Fig. 1 The overall vehicle control system architecture.

9.2.1 System functional control

The main objective of this control layer is to optimize the performance of the EV driveline subsystems while making sure that there is no functionality conflict or malfunctioning behavior between these local controllers. To achieve this goal, system function control algorithms includes two main control strategies as below:

- Controlling each local controller performance to achieve predefined functionalities.
- Manage the required and predefined functional relationships and interactions between the local controllers.

Controlling the local controllers is usually through CAN communication and can be as simple as a closed-loop On/Off control based on feedback for some subsystems such as air compressor On/Off control based on the air pressure feedback of front and rear brake lines or as complex as torque mapping strategies for propulsion motor control as discussed in the chapter on Drivetrain Control System (Chapter 7).

All HV subsystems must work in the safe and optimal condition, and performance of one of them must not create a negative impact on the other. An example of such a functional conflict that affects vehicle performance is torque mapping. Torque mapping is the process of defining the command torque to the motor controller based on the driver request. Battery pack, electric motor, and inverter status must be considered in torque mapping to avoid creating any hazard to the system. For example, if the motor or battery pack temperature is higher than the predefined safe limits, then commanding maximum available torque results in damaging the motor windings/permanent magnet rotor and battery cells accordingly; that will result in EV driveline breakdown. Another example could be regenerative braking control while maintaining efficient motor performance under various load demand.

In heavy-duty vehicles, the main system functional control tasks that need to be implemented in SC can be summarized as below:

- CAN control of the HV subsystems, such as motor controller, DC/DC converter, battery charger, HV-PDU, and steering system.
- Torque mapping strategies according to the drive demand, including required torque limiting and drivetrain degradation strategies aligned with HV subsystems status.
- Regenerative braking control under various load conditions.

- Controlling DC/DC converter output current and voltage based on LV load demand.
- Liquid cooling system control and monitoring for optimal performance of HV subsystems.
- Electric-driven air compressor and steering system control in different condition.

9.2.2 Vehicle operation mode control

Vehicle operation modes refer to various operating conditions that either is requested by the driver or decided by the SC according to the subsystem status. Ignition key position, gear position, accelerator/brake pedal, charging plug switch, and steering wheel angle are examples of driver inputs to the SC. In a simple word, SC decides the vehicle mode operation and hence the overall performance of the related subsystems.

The vehicle operation mode control is a layer above the system functional control, which can override the strategies within the functional control layer, according to the vehicle operation mode. In a different word, the subsystems functional control strategies are specific to each operation mode, and SC is responsible for implementing them accordingly. For instance, the vehicle cannot be driven in charging mode as the drivetrain is deactivated, while the HV battery is charging by the battery charger. Finite state machines (FSMs) are usually used for implementing EV operation mode control strategies in the SC.

Respective vehicle operation modes are dependent on the system design, overall vehicle control topology, and the application requirements. A typical EV operation modes model is shown in Fig. 2. The presented model is the most straightforward available architecture; however, this architecture could be much more complex and detailed based on the vehicle control system architecture.

There are two main streams of operation modes—normal operation and fault operation modes. Normal operation mode refers to the scenario that all subsystems are working as expected; all feedback parameters are within predefined limits, and there is no fault or warning associated with them. Fault mode refers to the scenario when the performance of one or more local controllers' is not idle, some feedback parameters are out of limits or there is a fault in the system. Therefore the SC is responsible for defining the correct state of operation of the EV according to the driver demands, EV driveline status, and diagnostics reporting.

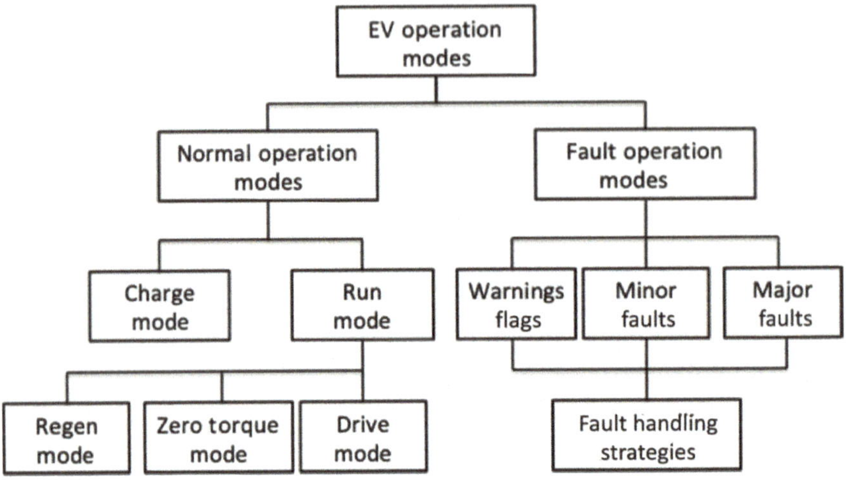

Fig. 2 A typical EV operation modes model.

9.2.3 Diagnostics and fault handling strategies

This control layer includes some of the most critical strategies and safety functions of the SC owing to existence of HV and high-power energy sources in EV applications. In the automotive industry, voltages above 60 V are considered as HV applications, and relevant regulations need to be followed in the electrical system design. Additionally, extensive diagnostics must be carried out in the SC software to monitor all HV subsystems, and the control system must be able to implement required fault-handling strategies to maintain safe operation of the vehicle.

Diagnostics must be performed continuously in the background. It must include checking of all safety-critical parameters of the HV subsystems and other conventional subsystems of the vehicle. Most of the HV subsystems communicate their operation status, including internal faults and warnings over the CAN Bus. The main objective of these continual communications and checks is to make sure the safety-critical parameters are within the accepted limits and ensure any warning or faulty condition is addressed accordingly and communicated to the vehicle driver.

As shown in Fig. 2, fault mode operations are categorized to warnings, minor faults, and major/severe faults by the supervisory controller. Warnings are defined when one or more parameters are approaching the critical limits but have not passed the critical limits yet. The vehicle can be operational when the warning flags are raised; however, some strategies such as limiting drivetrain power may be implemented according to the warning flag. For example, if the HV battery temperature warning flag is raised, limiting the drivetrain torque results in limiting the battery current and potentially helps in reducing the cell temperature. All warnings need to be communicated to the driver through a dashboard display. Minor faults are defined when a parameter exceeds a critical limit, or a subsystem is in a fault condition; however, it is not going to create a significant safety issue or cause vehicle breakdown. An example of such a fault could be HVAC or DC/DC converter system failure. Major faults are defined when there are safety risks or vehicle operation is not possible. Example of such faults could be a major failure in battery or drivetrain subsystems.

Fault-handling strategies will be highly dependent on the fault severity, HV subsystems internal safety features, and the overall HV electrical design. The main goal of fault-handling strategies is to maintain the best possible vehicle safe operations. Fault-handling strategies could include one or more of items below according to the fault severity level, faulty subsystem, and vehicle operation mode.

- Warning to the driver (WtD).
- Limiting drivetrain torque/power + WtD.
- WtD + shut down request from the driver (according to the vehicle operation manual).
- Controlled shutdown by the supervisory controller + WtD + warning to the passengers (WtP).
- Emergency shutdown + WtD + WtP.

9.2.4 Case study

Vehicle control system development is a complex task and highly depends on the drivetrain configuration, vehicle HV and LV electrical system, and operations requirements. Following case study tries to give a guideline on the vehicle control system design and required diagnostics. Let's consider the heavy-duty vehicle driveline system as shown in Fig. 3. This driveline includes major subsystems below:

- HV battery, including BMS.
- Onboard battery charger.
- HV power distribution unit.
- A single drivetrain motor and motor controller.
- DC/DC converter.
- Electric-driven air compressor and steering system.
- HMI—driver display.

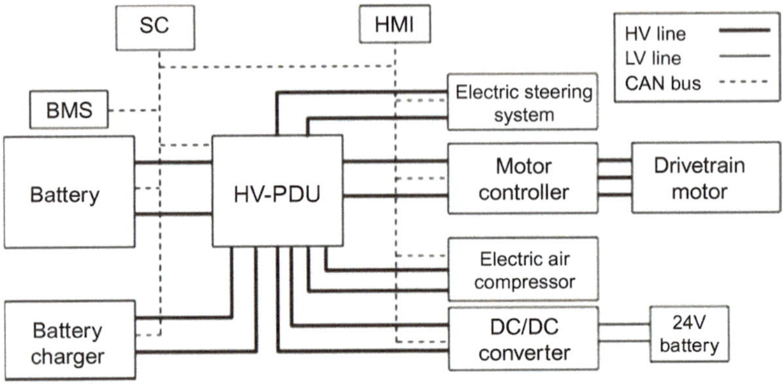

Fig. 3 Heavy-duty electric vehicle driveline system—case study.

The first step is to define all required CAN messages or physical signals of the subsystems and validate their performance and control options by the SC on the test bench. Then to define the normal operation modes based on the driver ignition key and charge plug inputs. Table 1 shows an example operation mode definition for our case study. Run mode and charge mode may have their internal operation modes accordingly.

Table 1 Normal operation modes—case study.

Normal operation mode	Ignition key position			Charger plug	Comments
	Off	Acc	On		
Off	1	0	0	0	Vehicle is Off. Charge is not active
Accessory mode	0	1	0	0	24V is connected to all subsystems. Vehicle cannot be driven. HV subsystems are not operational except DC/DC converter
Run mode	0	1	1	0	Vehicle is ready to be driven. All HV subsystems are operational. Charger plug is not connected
Charge mode	X	X	0	1	Ignition On position is not connected. Vehicle cannot be driven. Vehicle can be charged in Accessory Mode too

Fault diagnostics begins as soon as 24V is connected to the HV subsystems and CAN communications starts. The vehicle operation mode and diagnostic status are always shown on display for the driver. As discussed earlier, every subsystem sends their operation status and internal warning and fault flags over the CAN bus. As an example in this case study, the diagnostics parameters that we have are as below:

- **BMS**
  - Cell voltage (high, low, avg.).
  - Cell temperature (high, low, avg.).
  - Pack SoC and SoH.
  - Pack voltage.
  - Pack current.
  - Pack resistance.
  - Charge current limits.
  - Discharge current limits.
  - Internal warnings and fault flags.
- **PDU**
  - DC voltage.
  - DC current.
  - HV contactors status.
  - DC bus bar temperature.
  - Isolation measurement.
- **Drivetrain motor controller (same for air compressor and steering system motor controllers)**
  - Inverter state.
  - DC link power.
  - DC link voltage.
  - DC link current.
  - Maximum available torque.
  - Ref torque.
  - Mechanical power.
  - Motor speed/torque.
  - Phase current.
  - Motor temperature.
  - PCB temperature.
  - IGBT modules temperature.
  - System warnings/errors.
- **Battery charger**
  - DC output voltage.

  - DC output current.
  - AC input voltage.
  - AC input current.
  - AC power.
  - Charger temperature.
  - System warnings/errors.
- **DC/DC converter**
  - DC input voltage.
  - DC input current
  - PCB temperature.
  - IGBT modules temperature.
  - DC output voltage.
  - DC output current.
  - System warnings/errors.

The supervisory controller continuously monitors above diagnostic parameters and define the fault status of the vehicle as below:

- Severe (major) fault level.
- Medium (minor) fault level.
- Warnings.
- No fault (normal operation).

The list of possible faults is enormous, and Failure Mode and Effects Analysis (FMEA) document needs to be developed to define those faults, their risk level, and possible action to mitigate the effects. Going through a detailed FMEA development process is out of the scope of this book. In this section, significant/severe possible faults in an EV that needs immediate system shutdown are summarized as below:

- **Isolation failure**: Measured isolation resistance between HV terminals and vehicle chassis is below standard. More details are available in Chapter 5.
- **HVIL fault**: High Voltage Interlock Loop of one of the HV connectors is broken.
- **High G force fault**: Vehicle body experiences a G force higher than the expected value. Generally, it means an accident has happened.
- **High cell temp. fault**: The cell temperature goes above the upper critical limit.
- **Low cell temp. fault**: The cell temperature is below the lower critical limit.
- **High cell voltage fault**: The cell terminal voltage exceeds the upper critical limit.

- **Low cell voltage fault**: The measured cell voltage falls below the lower critical limit.
- **High motor temp. fault**: Motor temperature is above the critical limit.
- **Motor controller internal faults**: Including overcurrent, overvoltage, overtemperature, short circuit fault.
- **CAN time-out faults**: Loosing CAN communications with BMS, Motor controller, and Driver.

Customarily, the critical limits are available from product technical specification sheet, but lower values can be fed to the SC if a higher safety margin is required. SC is responsible for the safe operation of the vehicle, and any fault related to the HV battery system and drivetrain needs to be addressed immediately. Isolation failure and HVIL fault are highly critical, and emergency HV shutdown is required because it may lead to electrocution of passengers. Any severe mechanical shock or impact on the battery needs to be addressed by disconnecting the battery from the system immediately. Lithium-ion cells are highly sensitive to the operating temperature and their charge/discharge cell voltage and need to be protected against the extreme condition. Onboard HV battery is a huge source of energy, and any possible fault on cell performance can lead to a catastrophic failure and must be avoided. For example, limiting the drivetrain torque as discussed earlier, is one of the standard solutions. Drivetrain motor and its inverter drive also are part of critical subsystems, and their faults need to be categorized as high-priority. Any major electrical fault in the drivetrain may also result in battery failure. CAN communication is the backbone of the control system and needs to be monitored continuously. Any issue in the CAN bus or missing messages of major subsystem (BMS, HV PDU, and motor controller) are categorized as high-priority fault and results in the vehicle stop.

9.3 Power management strategies

As discussed earlier, the two main objectives of the SC are to maintain the performance of the EV within its safe operating limits and take appropriate actions in the event of a fault. Of these two, the performance of the EV heavily depends on the strategy used to manage instantaneous power demand and energy stored in the battery. The strategies used to manage these two aspects are referred to as power management strategies. Sometimes, the term "energy management strategies" has been used in the literature interchangeably to refer to the strategies mentioned earlier.

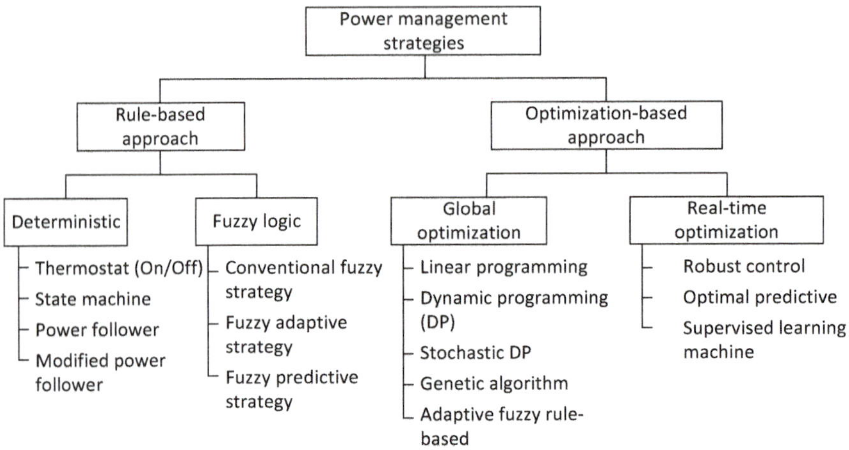

Fig. 4 Classification of EV power management strategies.

Power management strategies can be broadly classified into two groups as rule-based approaches and optimization-based approach. Rule-based approaches are primarily based on the rules derived from mathematical models, engineering knowledge, experience, intuition, predefined driving cycles, and load-leveling strategies of the EV. On the other hand, optimization-based approaches are based on analytical or numerical operations aimed at minimizing a cost function. Fig. 4 illustrates the classification of power management strategies developed for EVs.

Under each of the categories given in Fig. 4, a large number of power management strategies have been reported so far in the literature with the primary focus is on hybrid EVs. Nevertheless, because there is only one source of energy, which is the battery bank, in the heavy-duty EVs discussed in this book, some of the sophisticated power management strategies mentioned above are not required. In most cases, state machines with constraints and torque mapping techniques are sufficient for the fully battery powered EVs. Therefore power management aspects are not discussed in detail in this book. Interested readers can refer to [6–10] for more details on power management strategies applicable to the hybrid EVs.

9.4 Summary

EV supervisory controller structure has been presented, and details of system functional, vehicle operation mode, and fault and diagnostics control layers have been discussed in this chapter. Role of the SC in the safe operation of EVs has been highlighted and explained. Different SC internal control layers are also explained and discussed through an example case study for a better understanding.

References

[1] J.L. Bossa, G.A. Magallán, C.H. De Angelo, G.O. García, Implementation of a supervisory control system for an electric vehicle, in: 2010 9th IEEE/IAS International Conference on Industry Applications - INDUSCON 2010, Sao Paulo, 2010, pp. 1–5. doi:10.1109/INDUSCON.2010.5739893.

[2] R. Mura, V. Utkin, S. Onori, Energy management design in hybrid electric vehicles: a novel optimality and stability framework, IEEE Trans. Control Syst. Technol. 23 (4) (2015) 1307–1322, https://doi.org/10.1109/TCST.2014.2363617.

[3] C. Lin, H. Peng, J. Grizzle, J. Liu, M. Busdiecker, Control system development for an advanced-technology medium-duty hybrid electric truck, in: SAE Technical Paper 2003-01-3369, 2003. https://doi.org/10.4271/2003-01-3369.

[4] A.M. Phillips, M. Jankovic, K.E. Bailey, Vehicle system controller design for a hybrid electric vehicle, in: Proceedings of the 2000. IEEE International Conference on Control Applications. Conference Proceedings (Cat. No.00CH37162), Anchorage, AK, USA, 2000, pp. 297–302. doi:10.1109/CCA.2000.897440.

[5] V.K. Sharma, R.D. Kamble, R. Sharma, Control, alarm and indicator systems in modern electric vehicles, in: 2006 IEEE Conference on Electric and Hybrid Vehicles, Pune, 2006, pp. 1–5. doi:10.1109/ICEHV.2006.352290.

[6] S.F. Tie, C.W. Tan, A review of energy sources and energy management system in electric vehicles, Renew. Sust. Energ. Rev. 20 (2013) 82–102.

[7] A.M. Ali, D. Söffker, Towards optimal power management of hybrid electric vehicles in real-time: a review on methods, challenges, and state-of-the-art solutions, Energies 11 (2018) 476.

[8] B. Sakhdari, N.L. Azad, An optimal energy management system for battery electric vehicles, IFAC-PapersOnLine 48 (15) (2015) 86–92.

[9] H. Yu, Fuzzy logic energy management strategy based on genetic algorithm for plug-in hybrid electric vehicles, in: 2019 3rd Conference on Vehicle Control and Intelligence (CVCI), Hefei, China, 2019, pp. 1–5.

[10] H. Peng, J. Xie, Energy management strategy for plug-in hybrid electric vehicles based on genetic-fuzzy control strategy, in: 2018 International Computers, Signals and Systems Conference (ICOMSSC), Dalian, China, 2018, pp. 211–214.

CHAPTER 10

Technology Roadmap for Heavy-duty Electric Vehicles

10.1 Introduction

Along with the developments and increased penetration of clean energy technologies, heavy-duty electric vehicles (EVs) have become a promising solution to reduce emissions. Nevertheless, large-scale deployments of heavy-duty EVs, especially electric buses, have not been reported so far. This is mainly because of several challenges, which could be categories into three main groups, namely: technical challenges, financial barriers, and institutional barriers. Technical challenges are mainly associated with vehicle and battery technologies, lack of operational data, impacts on the grid, and limited availability of charging infrastructure. Besides, if the grid remains coal- or gas-powered, then it is hard to claim low or zero emissions as it shifts emissions from one locality to another rather than reduction or elimination. From the financial perspectives, barriers include the high initial cost of vehicles and battery systems, limited financial options, high risk associated with the investments, and large capital investments required for developing grid and charging infrastructure. Institutional barriers are mostly the communication gap between manufacturers and decision-makers, lack of a plan to remove existing vehicles, lack of policies on adopting EVs, and lack of spaces to install charging infrastructure.

The following sections discuss aforementioned challenges in detail. Subsequently, technological advancements and developments that are expected to support large-scale heavy-duty EV deployment are presented in the "technology roadmap" section.

10.2 Challenges to heavy-duty EV deployment

10.2.1 Technical challenges

10.2.1.1 Performance

It is common, and fair as well, to expect EVs to match, or in some cases exceed, the performance of corresponding heavy-duty diesel vehicles.

Heavy-duty Electric Vehicles
https://doi.org/10.1016/B978-0-12-818126-3.00011-7

In electric buses, the most common comparison is mileage. In some other types of heavy-duty EVs, such as earthmoving vehicles, performance comparisons could be in terms of power and work duration with a single charge. The mileage or work duration mainly depends on the energy capacity of the battery. Even though adding more battery modules can increase the energy capacity, it increases the weight and cost of the vehicle. Therefore batteries with high energy density are always preferred.

Moreover, the lifecycle cost of the battery is also a crucial factor. Thus the trends are to develop battery technologies that do not use or reduce the use of expensive metals such as cobalt [1]. Increasing the cycle life of batteries is also another trend in the battery industry because it effectively reduces the lifecycle cost of EV batteries. Weather dependency of battery performance is also a challenge, especially in extreme environments, which demands part of stored energy to be used for heating or cooling the battery. Therefore developing new battery technologies with extended operating temperature ranges has become the primary aim of many research programs worldwide.

10.2.1.2 Energy storage systems

The main objectives of battery research programs are to increase the energy density, power density, cycle life, charging rate, safety and temperature stability, and manufacturability, while reducing the cost [1,2]. Lithium (Li)-ion-based technologies are currently dominating the battery market with energy density in the range of 250 Wh/kg. The targets are to reach 350 Wh/kg by 2025 and 500 Wh/kg by 2030, which will effectively reduce the battery weight to a half. The life of Li-based batteries is closer 4000 cycles, and the targets are to achieve more than 5000 cycles by 2025 and reach 10,000 cycles by 2030 [3,4]. Advanced Li-based battery technologies such as Li-sulfur, nickel manganese cobalt oxide (NMC), and Li-air are the forerunners in the race to achieve these targets [1].

Moreover, non-Li-ion battery technologies such as zinc-air, sodium-ion, and aluminium-ion batteries are actively being researched, aiming to meet the above performance targets. Voltage, current, and temperature measurements at the cell level and use of Artificial Intelligence (AI) techniques to estimate the State of Health (SoH) are other prominent trends in battery research. These technologies help to manage the battery efficiently and extend the cycle life.

10.2.1.3 Inverter drives

The drivetrain is the other crucial part that directly affects the energy efficiency of EVs. The conversion of electrical energy stored in the battery into mechanical energy should occur with minimal losses in a wide range of speeds. Therefore the power electronic inverter, which converts the battery DC voltage into the required AC voltage and frequency of the motor, should maintain its performance in a wide operating range. Therefore the trend is to develop power electronic converters with advanced switching devices such as silicon carbide (SiC) to obtain low switching losses and low on-state resistance over a wide operating temperature range.

The requirements of the power density of power electronic converter are not very stringent in heavy-duty EV when compared to electric cars because they are relatively big and heavy. Nevertheless, weight and volume savings that can be achieved through high-density power converters are always beneficial. The current EV inverter technology is in the range of 30 kVA/L in terms of power density, and the target is to achieve 100 kVA/L with advanced packaging using wide bandgap devices such as SiC and gallium nitride (GaN) [5,6]. Two-level voltage source inverter topology will be sufficient to meet the power levels of heavy-duty EVs. Nevertheless, multiphase two-level inverter topologies can be used for applications that require high torque [5].

Thermal management becomes challenging when the power density increases. Passive two-phase immersion cooling technologies are gaining attention for efficient heat removal. Other trends in cooling technologies include jet impingement and spray cooling, which have received significant research attention in recent years, along with microchannel heatsinks and heat pipes.

10.2.1.4 Electric motors

In addition to the power electronic inverters, recent developments in drivetrain electric motor technologies have introduced axial flux motors and switch reluctance motors to enhance drivetrain performance in wide operating ranges [7]. Axial flux motors have an axial air gap, rather than conventional radial flux motors with the radial air gap. Axial flux motors produce a higher torque when compared to radial flux motors of the same size. Therefore they have been of interest of EV developers in recent years because of their high power to size ratio.

10.2.1.5 Charging

Availability of charging infrastructure has always been identified as an essential aspect to promote the widespread use of heavy-duty EVs because it would help to develop user confidence. Nevertheless, the high initial cost of such infrastructure is a significant barrier. Besides, large-scale penetration of EVs has a negative impact on the grid because it could lead to imbalances in demand and supply [8]. Therefore instead of opportunity charging, scheduled charging with optimal allocations are being recommended and often provided with economic incentives.

10.2.1.6 Policy makers

Apart from the abovementioned technological challenges, lack of knowledge among decision-makers on heavy-duty EV technologies and advancements has been identified as a barrier. Specifically, the information on operational characteristics, limitations, and maintenance requirements of heavy-duty EVs available on the market are to be readily available for decision-makers.

10.2.2 Financial barriers

A high initial cost is the primary challenge in procuring heavy-duty EVs. For example, an electric bus would cost two or three times more when compared to a conventional diesel bus. From a business perspective, this high initial cost is a result of being a new technology associated with unknown risks. These risks come from the uncertainties on long-term battery performance, maintenance requirements, and the residual value. Also, large-scale production has not been established yet for heavy-duty EVs, and thus benefits related to the economics of scale are not available to the investors. Even though heavy-duty EVs will become price-competitive over time, it is unclear when they will reach that stage [9].

Lack of financing options at all stages of heavy-duty EV adoption is another barrier. As a solution, manufacturers can offer leasing options [9]. Moreover, manufacturers can provide favorable leasing to potential owners one of their manufacturing plants because it reduces transport and maintenance costs. New forms of financing, such as "pay-as-you-save" and separate battery leasing, are also being proposed as solutions.

10.2.3 Institutional barriers

In the case of electric buses, it has been hard for some cities to implement electric bus programs because of barriers between manufacturers and transit agencies. The geographical distance between the manufacturer and the city where the electric buses are to be deployed has negatively impacted implementation plans because of delays in procurement and delivery. Tariffs and bureaucracy associated with international trade have been identified as other challenges when procuring electric buses from another country. When electric buses require spare parts for maintenance, they have to be shipped from the manufacturing plant, which always linked to delays, loss of revenue, and disruption to the service [9].

As a solution, some manufacturers have started to build manufacturing plans closer to the potential markets. Lack of a plan to remove the current heavy-duty vehicles would also be a practical challenge and sometimes associated with contractual lock-ins. As a result, the mass adoption of heavy-duty EVs could be delayed. Moreover, lack of space and land to install infrastructure and little planning for long-term implications in the grid have been reported as other institutional barriers for adopting heavy-duty EVs [9]. This could be addressed through long-term planning and discussions involving relevant institutions.

10.3 Technology roadmap

10.3.1 Energy storage system

Improvement in EV technology will naturally stem from the developments made in the field of batteries. It does not come as a surprise that 95% of all electric buses manufactured in China in the year 2018 had lithium iron phosphate (LFP) battery packs. LFP battery packs are inherently safer than other chemistries such as Li nickel cobalt aluminium oxide (NCA) and NMC. Therefore LFP batteries have been promoted extensively through various subsidies for use in EVs despite their much lower energy densities. Increasing demand for high-performance vehicles is making the market move away from the traditional choice [10].

Battery packs for the next-generation electric trucks and buses should be lightweight and extremely compact, that is, they should possess high energy and power densities to maximize the usable space and energy efficiency of the system. They must also demonstrate a reliable cycle life and reduction in overall cost of the battery system [11,12]. Therefore nickel content in

Battery generation	Chemistry	Commercialization
Generation 5	Lithium-Air batteries	15 years and beyond
Generation 4b	Solid-state batteries with Li metal anode and conversion materials, for example, Li Sulfur as cathodes	8–15 years
Generation 4a	Solid state batteries with graphite anode and conversion materials as cathodes	5–8 years
Generation 3b	Cathode: High voltage spinel materials, for example, LNMO, disordered rock salt; Anode: silicon/carbon	2–5 years
Generation 3a	Cathode: NMC622, NMC811; Anode: Graphite + silicon (5%-10%), LTO	State-of-the-art (year 2020)
Generation 2b	Cathode: NMC523, NMC622; Anode: Graphite, LTO	
Generation 2a	Cathode: NMC111; Anode: Carbon (100%)	
Generation 1	Cathode: NCA, LFP; Anode: Carbon (100%)	

Fig. 1 Timeline for anticipated commercialization of different generations of Li-ion batteries.

third-generation Li-ion batteries is increasing sharply, whereas cobalt content is decreasing. Chapter 4 presents a comprehensive overview of the advancements made in the material domain to improve the performance of the state-of-the-art Li-ion batteries.

However, Li-ion batteries have certain limitations by virtue of their design and chemistry, which cannot be overcome. Hence novel battery concepts such as solid-state batteries and metal-air batteries are being pursued. Fig. 1 presents an optimistic time frame by which each of the next-generation batteries may become commercially viable. The following text offers some description of the existing challenges.

10.3.1.1 Solid-state batteries

The specific energy density of the state-of-the-art Li-ion batteries is approximately one order of magnitude smaller than the specific energy density of petrol [13]. Plus, they contain a liquid electrolyte to facilitate charge transfer. Liquid electrolytes have high flammability, which is a safety hazard. In comparison, electrolytes that exist in a solid state are much safer and enable higher gravimetric and volumetric densities [14,15]. This is evident from the illustration presented in Fig. 2. Therefore research is underway to find suitable solid alternatives for the next-generation Li-ion batteries. The potential candidates must demonstrate high ionic conductivity, low diffusion coefficient for species such as H_2O and O_2, exhibit excellent interfacial compatibility with both the anode and the cathode, that is, high resistance to any side reactions with Li metal, and must have high manufacturability.

Solid-state batteries can be designed with a graphite anode or with another novel anode material. Noteworthy is that benefits gained in

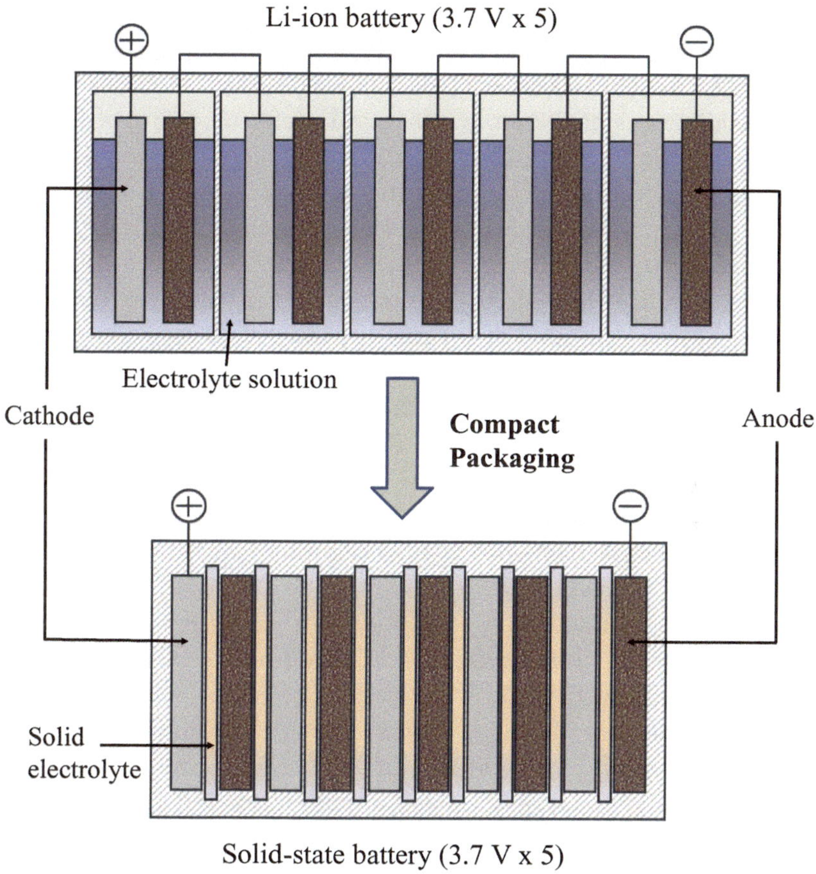

Fig. 2 Comparison of volumetric densities of Li-ion battery and all-solid-state battery. Directly connected battery cells in all-solid-state battery enables smaller and tighter packaging.

terms of cost savings and reduced packaging volume because of a higher energy density of solid-state battery are only incremental for the cases when the graphite anodes are used. In contrast, the highest packing density of Li atoms and, consequentially, ultrahigh theoretical specific capacity are achieved by replacing the graphite anode with a metallic Li anode (3860 mAh/g vs. 372 mAh/g). Additionally, metallic Li anodes illustrate the lowest electrochemical potential (−3.04 V vs. the standard hydrogen electrode) and a low density in comparison to graphite (0.59 g/cm^3 vs. 2.26 g/cm^3) [16]. However, Li metal anodes are prone to dendrite formation, which has been the main stumbling block.

During battery cycling, Li is electrochemically stripped from one electrode and plated on to another. The process continues as the battery is repeatedly charged and discharged. The issue is that the electrochemical plating process does not result in a nice and uniform coating of Li on the counter electrode "per se"; instead, it initiates nucleation and growth of Li dendrite projections. The dendritic growth causes a reduction in columbic efficiency and leads to irreversible capacity loss. If the growth is left unchecked, the dendrites are also capable of causing internal short circuit by piercing through the separator. It can result in the battery being pushed into the state of thermal runaway. This is an unacceptable safety hazard. Naturally, substantive research efforts are being directed toward developing a range of methods to suppress dendrite growth. Promising approaches include:

1. Adding electrolyte additives to promote stable solid-electrolyte interphase (SEI) growth [17,18].
2. Creating artificial, stable SEI layer at the interphase of electrolyte with the Li metal anode [19,20].
3. Adopting current collectors and anodes with a 3D porous structure in cell assembly to homogenize Li-ion flux, thereby regulating Li-ion deposition process [21,22].

Limited success has been achieved though, mainly because the electrochemical plating reaction kinetics favor Li dendrite nucleation and growth. Recently, Lu et al. disclosed a method for self-heating-induced healing of Li metal dendrites in an international patent WO 2019/191530. According to this, the battery is cycled at healing current that equates to a current density greater than the operating current density. Operating current density refers to a standard or a predefined current value that results in optimum performance and durability for the target application. Generally, it has been noticed that operating batteries at high current rates accelerate formation of metal dendrites over the anode. However, Lu et al. demonstrated that if the operating current density is increased sufficiently, the heat produced within the battery triggers a self-healing process. It ensures a smooth anode surface that is much less likely to cause an electrical short circuit. The generated local heat flux disturbs the system equilibrium and initiates random flow of Li atoms making the dendrite. The diffusion of Li metal particles from the dendrite surface smoothens the anode as the adjacent dendrites fuse together, thus resulting in anodes with a flat surface configuration. It must be emphasized that the healing current is always restricted below the manufacturer's specified maximum permissible value. Therefore there is minimal risk of any thermal damage being caused to the separator or electrolyte breaking down because of the high currents used in the healing process [23].

This is one step in the right direction. To successfully commercialize this technology, several other challenges need to be addressed. For example, batteries must be designed for recyclability—Li-ion batteries are not designed for recycling. Consequentially, valuable materials, including Li, nickel, and cobalt, are lost when the batteries are disposed at the end of their life. It is, therefore, recommended that a sustainable lifecycle approach must be adopted for next-generation energy storage devices such as the solid-state batteries that are yet to reach commercial deployment phase in their development cycle [14]. Appropriate recycling strategies must be developed beforehand to implement the "production to recycling" sustainable manufacturing philosophy. National and international regulations are needed to drive research, design, and development efforts at various levels in this area [24].

Last, the high price of solid-state batteries is another issue. It is expected that cost-competitive solutions that can challenge the incumbent Li-ion battery technology will become available after 5–10 years at least.

10.3.1.2 Lithium-air batteries

Metal-air batteries, for example, zinc-air battery, aluminium-air battery, and Li-air battery, as the name suggests, use pure metal as the anode in place of an intercalation or conversion anode material used in Li-ion batteries. Theoretically, the choice of anode material is the only factor that influences their energy density. Because Li is the lightest element, the theoretical energy density of the Li–air battery is the closest to that of petrol (11,680 Wh/kg vs. 13,000 Wh/kg). Corresponding to a potential of 3.0 V, it equates to a theoretical capacity of 3862 Ah/kg, which is a 10-fold increase over the theoretical capacity of conventional Li-ion batteries at the cell level. If the weight contribution of inactive components is considered, then also there is at least three to four times gain in the systems level energy density over Li-ion batteries [25–27]. Unfortunately, it has been a significant challenge to translate these benefits into an operational system.

Based on the type of electrolyte used, Li-air batteries can be categorized as—nonaqueous batteries, aqueous batteries, hybrid batteries, or solid-state batteries [28]. Fig. 3 illustrates the difference in their design. Essentially, they are an example of an open system because their positive electrodes are open to the environment and participate directly in inexhaustible oxygen

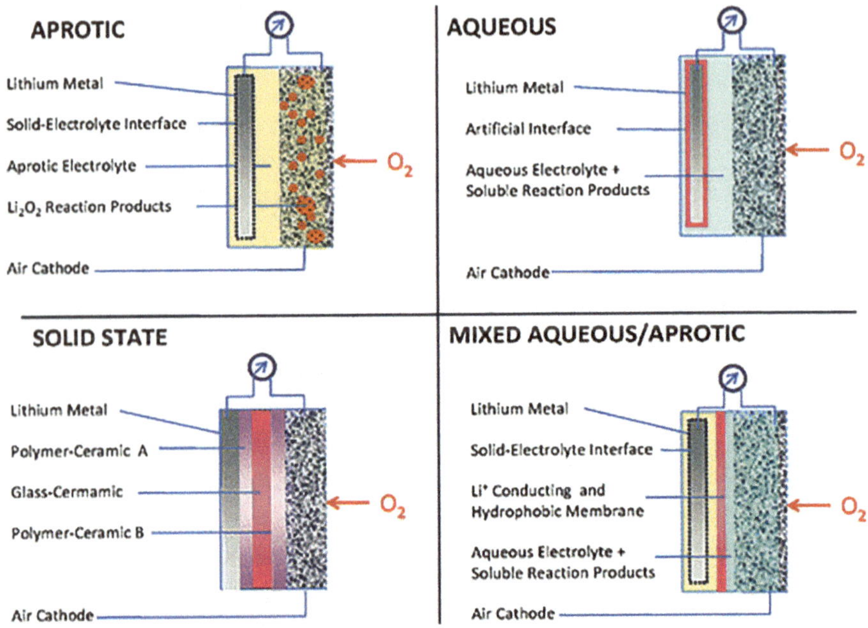

Fig. 3 Four different architectures of Li-air batteries. Principal components are as labeled in the figure. Spontaneously occurring SEIs on the Li metal anode are given as dashed lines, while artificial SEIs are given as solid lines. *Reprinted with permission from [28]. Copyright (2010) American Chemical Society.*

exchange with the ambient air. The oxidative agents present in the air such as carbon dioxide and moisture contaminate the metal anode and deteriorate its performance [29].

Furthermore, the oxygen reaction kinetics, that is, oxygen evolution reaction (OER) and oxygen reduction reaction (ORR), characterizing Li-air batteries, are sluggish. It results in a poor round-trip efficiency, low charge/discharge capacities, and short cycle life. Subpar system design and a lack of complete understanding of the battery electrochemistry that may sometimes get termed as "underperforming battery materials" render the Li-air battery technology commercially unviable at this time.

Basic electrochemistry of a Li-air battery is defined by Eq. (1):

$$2Li + O_2 \leftrightarrow Li_2O_2 \tag{1}$$

Theoretically, the quantity of Li_2O_2, that is, the discharge product produced on the air cathode determines the overall discharge capacity or energy density of the battery. Therefore a porous electrode structure offering a large reaction area is used for constructing the batteries.

The porous structure provides a network of diffusion channels that facilitate efficient oxygen transport through to the electrode-electrolyte interface. The discharge product exists in a solid phase in the nonaqueous and the solid-state Li-air batteries. The porous design provides voids within the electrode structure to accommodate the solid Li_2O_2. Over time, these deposits start blocking the pores and create clogging [30]. Clogging restricts the airflow in addition to affecting Li-ion and electron transfer, which in turn causes the practically available discharge capacity of the Li-air batteries to become significantly lower than the theoretical values. Thus the porosity features such as pore size and pore volume need to be carefully engineered to realize high capacities.

Owing to a limited oxygen transport rate through the porous cathode, the concentration of oxygen varies across its thickness. Typically, a lower concentration is expected near the separator than in the region that is closer to the side that is open to the environment. Lower oxygen concentration effectively means lower Li_2O_2 production reaction rates near battery core. Consequentially, pores on the outer side get clogged first. Blockage of the air/oxygen pathways result in premature termination of the discharge process, leading to smaller than expected discharge capacities.

For the same reason, void volume and reaction side utilization is relatively low in batteries with thick electrodes, which explains their low specific capacity. It is important to mention here that the void volume utilization is affected by the morphology of the deposited product as well. The solid Li_2O_2 can exist in a particle-like form or can have a disc-like or a toroidal or any irregular shape. It can also have a thin film-like morphology. If the pore structure remains the same, deposits with a particle-like morphology have a higher void volume utilization than the film-like deposits. Donor number (DN) of solvents affects the morphology of Li_2O_2. In addition, operating conditions can influence the product morphology as well. For example, film-like Li_2O_2 is formed at low potentials in a solvent with low DN, whereas large Li_2O_2 particles are deposited in high DN solvents at high potentials [31].

The round-trip efficiency, on the other hand, is mainly dependent on the catalyst activities during the oxygen evolution and the ORRs. Li_2O_2 and the catalyst are both solid products. Solid Li_2O_2 covers the catalyst surface partly and blocks the transport of reactants. Catalytic mechanisms at the solid-solid interface are a limiting factor in the oxidation kinetics of the Li-air batteries. Unsatisfactory catalytic activity in OERs is responsible for a higher charge potential, whereas insufficient activity during ORRs results in a lower discharge potential. Poor electric conductivity further increases the electrode overpotential. The resultant widening of the gap between the charge and discharge voltages marginalizes the round-trip efficiency. This can be improved through proper electrode engineering as well.

Rechargeability or cycle life of the Li-air battery is determined by the electrochemical reversibility of the discharge product(s) within the operating voltage range of the battery's air cathode. Because of the open system design, the fundamental electrochemistry is affected by the contaminants such as moisture and carbon dioxide. In general, the average OER decreases with the increasing concentration of moisture (H_2O) in the air flowing into the battery. Dramatic reductions in cycling performance have been noted for battery operation in environments with a relative humidity greater than 15%. Furthermore, the H_2O reacts with Li_2O_2 to form LiOH.

Similarly, Li_2O_2 and CO_2 combine to form Li_2CO_3. Formation of these byproducts compromises the discharge voltage and discharge capacity. Furthermore, higher potentials are required to initiate their decomposition during the charge step, which influences the energy efficiency of the process and the cycling stability [32].

The open system design of Li-air batteries poses another significant issue. As the positive electrode of the battery is exposed to the ambient environment, the electrolyte in its vicinity evaporates. Decreased amount of electrolyte is correlated to the reduced interfacial region between the electrode and the electrolyte. The effect is observable in the form of slow charge transfer kinetics and reduced cycle life. US Patent 9263779 discloses a battery architecture that can prevent electrolyte in the Li-air battery modules from getting exhausted due to evaporation of the solvent. Herein, the battery modules are arranged in a housing made from an insulator material. The housing separates the positive and the negative electrodes in the battery electrically.

Furthermore, the housing includes an inlet for injecting the fresh air in to and an outlet for discharging the used air out of the housing. This selective injection limits the communication of battery cells with the outside environment. To maintain electrolyte level at or near saturation state in the battery module, additional electrolyte is stored in an accommodating unit, that is, a container of suitable size placed appropriately inside the housing. In addition, a collector or a cooling mechanism is installed in the path of the discharged air near the outlet. As a result of cooling, the electrolyte in the exiting air condenses and is collected in the accommodating unit.

It is envisaged that electrolyte evaporation will be further suppressed if the chosen electrolyte is composed of a Li salt, a Li-ion conductive polymer and a solvent that is a compound chemically represented by the formula:

$$R_1 - O - (\underset{R_3}{\overset{R_2}{\overset{|}{\underset{|}{C}}}} - \underset{R_5}{\overset{R_4}{\overset{|}{\underset{|}{C}}}} - O)_n - R_6$$

wherein each member of the *R* group is independently selected from a range of specific compound groups such as the nitro group, the carboxyl group, substituted or unsubstituted aryl and alkyl groups, a hydrogen atom, or a halogen atom. The Li-ion conductive polymer and the solvent compound are selected such that they have similar structural units. When combined, they react strongly with each other and provide a reduction in the volume of the evaporated solvent [33].

10.3.2 Vehicle systems

Through the accelerated rollout of renewable energy-powered electric grids and expected improvements in traction motors, battery packs, and their manufacturing methods, the so-called "green" advantage and the environmental impact of heavy-duty EVs will increase. Going forward, a modular or reconfigurable driveline system, a lightweight chassis, and superior system integration, all will contribute to a reduced total cost of ownership (TCO) and return on investment period. The TCO includes asset acquisition cost, operating cost, and asset disposal cost. The operating cost of a vehicle primarily consists of the energy cost and the maintenance cost of the vehicle. It is understood that energy cost of an EV is dependent on the drive cycle (e.g., vehicle speed, gradient), driver behavior (e.g., rate of braking, acceleration), and auxiliary loads (e.g., air conditioning load). In the case of a bus fleet, operating costs also include other expenses such as costs for ticketing and fleet management.

Table 1 presents a cost comparison prediction for years 2020, 2030, 2040, and 2050 with the TCO for different vehicle categories in the year 2017. Costs calculated are in USD. It is expected that the TCO will decrease markedly in the trucking sector between the years 2020 and 2050 when compared to the reduction estimated for the transit buses and the intercity electric buses. The difference can be attributed to a disparity in operating costs.

Table 1 Projected cost comparison for different heavy-duty electric vehicle categories based on the information available from [34].

	Vehicle weight (kg)	Driving range (km)	Cost (k$ - USD value in 2017)				Mileage (kWh/km)	Break-even kms in 2030
			2020	2030	2040	2050		
Transit bus (325 kWh, 250 kW)	13,750	240	509	443	437	429	1.35	168,000
Intercity bus (500 kWh, 250 kW)	14,850	560	616	489	471	434	0.89	422,400
HD pick-up truck (80 kWh, 225 kW)	3950	240	99	59	56	50	0.34	158,400
City delivery truck (150 kWh, 150 kW)	6900	240	113	79	75	66	0.625	64,000
Short-haul truck (350 kWh, 300 kW)	20,750	240	261	175	162	140	1.46	286,400
Long-haul truck (900 kWh, 350 kW)	29,500	480	389	213	194	169	1.875	603,200

10.3.2.1 Autonomous electric vehicles

Statistics indicate that 94% of serious road accidents are caused by lapse in concentration or false judgment of the human driver [35]. Human error is removed from the crash equation through deployment of autonomous vehicles (AVs) on roads. Omission of human factor from driving can reduce the incidents of road crashes, thereby minimizing the injuries caused to vehicle passengers and other road users. SAE standard J3016 describes taxonomy of AV categories. These are graphically shown in Fig. 4. The six categories range from no automation or level 0 to full automation or level 5. They are based on the number of automation features engaged in dynamic driving tasks at any given instance. Please note active safety systems, such as electronic stability control and automated emergency braking, and certain types of driver assistance systems, such as lane-keeping assistance, are excluded from the scope because they provide only momentary intervention during safety situations.

Level 0 —no automation, that is, the human driver performs all the driving-related tasks. There is zero autonomy.

Level 1 —driver assistance. The driver controls the vehicle, but some assistance may be available, that is, the advanced driver assistance system (ADAS) is included in the vehicle design which can help either in acceleration and braking or in steering the vehicle. It cannot control both operations simultaneously.

Level 2 —partial automation; herein, the ADAS of the vehicle can control both the steering and the acceleration/braking functions simultaneously. However, the driver must provide continuous monitoring of the driving environment and must remain engaged with other driving-related tasks.

Level 3 —conditional automation—at this level of autonomy, the automated driving system (ADS) can drive the vehicle without continuous monitoring from the human rider in some conditions, but s/he must remain available to take control at a moment's notice.

Level 4 —high automation. Level 4 automation refers to the fact the human driver is optional, that is, can choose to give up the controls of the vehicle because the ADAS can perform driving-related tasks with minimal supervision under certain conditions. The human driver is not required to pay attention under those conditions.

Level 5 —full automation. A fully automated vehicle can drive itself autonomously under all conditions. The humans riding the vehicle are simply passengers with no involvement in driving-related functions [36].

Autonomy Level 0	Autonomy Level 1	Autonomy Level 2	Autonomy Level 3	Autonomy Level 4	Autonomy Level 5
The human in driver's seat is **considered as driving** when the driver support features are engaged—regardless of feet position and any active steering control by the human driver			Even if seated in the driver's seat, the human driver is **regarded as NOT driving** when the automated support features are engaged		
To maintain safety, the human driver must constantly supervise operation of these support features			When the system requests, the human must take control of driving	These are fully autonomous features; the human passenger is not required to drive at any stage	
Driver Support Features			**Automated Driving Features**		
Level 0 features provide warnings and momentary assistance	These provide either steering or brake/acceleration support to the human driver	At level 2, support for both steering and brake/acceleration is available to the human driver	These provide support to the human drier under limited conditions, i.e., will not activate unless all required conditions are fulfilled		Level 5 features are fully capable of safe driving in all conditions
E.g., blind spot warning; lane departure warning; and automatic emergency braking	E.g., adaptive cruise control OR lane centering	E.g., simultaneous adaptive cruise control AND lane centering	E.g., traffic jam chauffeur	E.g., local driverless taxi-pedals/steering may or may not be installed	E.g., same as level 4—but can drive anywhere in any and all conditions

Fig. 4 Level of vehicle autonomy according to the SAE International standard J3016. (Color code: *Yellow* represents that human driver is considered as driving, while *green* represents autonomous operation when the support systems are engaged. Both the decreasing shade of *yellow* and the increasing shade of *green* indicate increasing level of vehicle autonomy).

The three actors in driving are: (1) the human user; (2) the ADS; and (3) other vehicles. Fully AVs that operate independently of a human driver are expected to contribute several folds to the increased passenger safety and better traffic management. In the case of autonomous electric vehicles (AEVs), efficient energy utilization by the interconnected fleet is an added benefit.

Some studies predict that small fleets of passenger AVs can replace public transit system [37–39]. However, integrating a fully autonomous driving system in vehicle design adds substantially to the TCO. Moreover, replacing buses with passenger cars would overburden our cities and highways, leading to congestion on roads. In addition, there is not much practical sense in developing fully autonomous fleets of passenger cars using the current technology. Heavy-duty vehicle sector, on the other hand, presents a compelling business case. Theoretically, the passenger-carrying capacity of an autonomous bus is greater than a regular bus of the same size because of the removal of the driver. Frequency of trips per bus can also be higher because the mandatory rest periods included originally to minimize driver's fatigue can be removed.

Nevertheless, technological complexity created by factors such as cross traffic at intersections, unpredictable movements of pedestrians and cyclists, and unreliable digital connectivity in specific regions make designing a fully autonomous bus extremely challenging. As far as autonomous trucking goes, it is anticipated to change the cost and utilization scale of the freight transportation sector. For fully autonomous trucks in the United States, operating costs would be up to 45% lower than the costs for the trucks with zero autonomy [40]. Tremendous improvements are, thus, being made in the field, and some companies have already started retrofitting medium and heavy-duty trucks with sensors, Light Detection and Ranging sensors (LiDAR), and other necessary technology to achieve some level of autonomous driving. Basic retrofitting kits are available for USD 30,000, whereas more advanced and high-end packages are available for USD 100,000.

Commercial deployment of autonomous trucks is expected to happen in three phases. Timewise, the first phase will go on at least until the year 2025 and will feature "platooning" of trucks. Platooning refers to connecting a convoy of trucks to the lead truck wirelessly. In the initial years of the first wave though, the convoy will consist of two SAE level 3 AVs only, which use computational algorithms to establish a link. Furthermore, platooning will take place on interstate highways, and a human driver's presence will be essential in each truck. As the technology matures in the later period, a driver in the lead truck will suffice to supervise the platooning operation of the unmanned truck following close behind. However, a driver in each truck would still be needed to drive on noninterstate highways. During phase one, the TCO will reduce by 10% from the 2018 levels. The subsequent phase will see the deployment of SAE level 4 AVs on the road. Convoys of unmanned trucks will ply themselves with constrained autonomy on interstate highways. Drivers will meet the trucks at dedicated exit stops and drive them to their final destination inside the cities. Essentially, the number of unmanned trucks operating in "geofenced" areas and on interstate highways outside of a platoon will increase a minimum of a few hundred units in the 2030s. However, their autonomous operation will still be subjected to acceptable communication with the traffic management system, visibility, and other weather conditions. Furthermore, fully autonomous trucking at large scale will be visible from the loading point till final delivery stations in the 2040s. Seemingly, SAE level 5 autonomous trucks will replace the fleet of traditional trucks widely, and the TCO will decrease by 45% in comparison to the 2018 marker [41–43].

The increasing level of vehicle autonomy could, however, have a negative impact on grid storage capabilities. Shared AVs would mean that each vehicle's battery bank would be connected to the electric grid for a far less time than initially anticipated. The causality must, therefore, be carefully investigated while calculating the grid storage facility size as well.

10.4 Summary

This chapter aims to support its readers in developing an understanding of various conflicting forces at work in the field of heavy-duty EVs. First, barriers such as technical, financial, and institutional barriers to the development and large-scale deployment of heavy-duty EVs are summarized. Then, technology advancements that are being pursued to support the ambitious goal of making heavy-duty EVs a cost-competitive and a commercially successful transport solution are presented. It is envisaged that majority of the success will come in response to the technological leap made in the energy storage domain in the form of solid-state battery technology and Li-air battery system and at the vehicular level through driverless or autonomous driving systems. This chapter offers a brief description of various aspects of these breakthrough technologies.

References

[1] M.K. Loganathan, C.M. Tan, B. Mishra, T.A.M. Msagati, L.W. Snyman, Review and selection of advanced battery technologies for post 2020 era electric vehicles, in: 2019 IEEE Transportation Electrification Conference (ITEC-India), 2019.

[2] Technical report of the Battery 2030+Roadmap program: Inventing the batteries of the future, research needs and future actions, 2020. March. Available online at https://battery2030.eu/digitalAssets/860/c_860904-l_1-k_roadmap-27-march.pdf.

[3] C. Aoxia, P.K. Sen, Advancement in battery technology: a state-of-the-art review, in: 2016 IEEE Industry Applications Society Annual Meeting, 2016.

[4] M. Meeus, Review of status of the main chemistries for the EV market, 2018, January. Available online at https://iea.blob.core.windows.net/assets/imports/events/71/Session1MeeusSustesco.

[5] J. Reimers, L. Dorn-Gomba, C. Mak, A. Emadi, Automotive traction inverters: current status and future trends, IEEE Trans. Veh. Technol. 68 (4) (2019) 3337–3350.

[6] B. Şarlioğlu, Automotive power electronics: current status and future trends, in: 2019 International Aegean Conference on Electrical Machines and Power Electronics (ACEMP) & 2019 International Conference on Optimization of Electrical and Electronic Equipment (OPTIM), 2019.

[7] M.Y. Metwly, M.S. Abdel-Majeed, A.S. Abdel-Khalik, R.A. Hamdy, M.S. Hamad, S. Ahmed, A review of integrated on-board Ev battery chargers: advanced topologies, recent developments and optimal selection of Fscw slot/pole combination, IEEE Access 8 (2020) 85216–85242.

[8] H. Patil, V.N. Kalkhambkar, Grid integration of electric vehicles for economic benefits: a review, J. Modern Power Syst. Clean Energy (2020) 1–14, https://doi.org/10.35833/MPCE.2019.000326.

[9] R.G.C. Sclar, S. Castellanos, X. Li, Barriers to Adopting Eletric Buses, World Resources Institute (WRI) Publication, 2019. May. Available online at https://www.wri.org/publication/barriers-adopting-electric-buses.

[10] C. Bos, How Catl Lithium-Iron-Phosphate Batteries Could Be Leading to 100 Kwh Tesla Model 3, 2020, Available from: https://cleantechnica.com/2020/02/18/how-catl-lithium-iron-phosphate-batteries-could-be-leading-to-100-kwh-tesla-model-3/. (Accessed 11 August 2020).

[11] S. Arora, A. Kapoor, W. Shen, Application of robust design methodology to battery packs for electric vehicles: identification of critical technical requirements for modular architecture, Batteries 4 (3) (2018) 30.

[12] S. Arora, W. Shen, A. Kapoor, Designing a Robust Battery Pack for Electric Vehicles Using a Modified Parameter Diagram; SAE Technical Paper 2015-01-0041, SAE International, Melbourne, Australia, 2015, https://doi.org/10.4271/2015-01-0041.

[13] P. Tan, H.R. Jiang, X.B. Zhu, L. An, C.Y. Jung, M.C. Wu, L. Shi, W. Shyy, T.S. Zhao, Advances and challenges in lithium-air batteries, Appl. Energy 204 (2017) 780–806.

[14] A. Gocha, Roadmap to Commercialize All-Solid-State Batteries, 2020, Available from: https://ceramics.org/ceramic-tech-today/energy-1/roadmap-to-commercialize-all-solid-state-batteries. (Accessed 3 September 2020).

[15] M. Neilson, Commercialising Solid-State Battery Technology—Are We at the Cusp of Great Things?, IP INFORMATION & NEWS, 2020. Available from: https://www.udl.co.uk/insights/commercialising-solid-state-battery-technology. (Accessed 28 September 2020).

[16] P. Beck, Solid-State Battery Tech: What's Close to Commercialization, and What's Still Years Away?, Tech Features, 2020. Available from: https://chargedevs.com/features/solid-state-battery-tech-whats-close-to-commercialization-and-whats-still-years-away/. (Accessed 28 September 2020).

[17] C.-Z. Zhao, X.-B. Cheng, R. Zhang, H.-J. Peng, J.-Q. Huang, R. Ran, Z.-H. Huang, F. Wei, Q. Zhang, Li2s5-based ternary-salt electrolyte for robust lithium metal anode, Energy Storage Mater. 3 (2016) 77–84.
[18] W. Li, H. Yao, K. Yan, G. Zheng, Z. Liang, Y.-M. Chiang, Y. Cui, The synergetic effect of lithium polysulfide and lithium nitrate to prevent lithium dendrite growth, Nat. Commun. 6 (1) (2015) 7436.
[19] R. Xu, X.-Q. Zhang, X.-B. Cheng, H.-J. Peng, C.-Z. Zhao, C. Yan, J.-Q. Huang, Artificial soft–rigid protective layer for dendrite-free lithium metal anode, Adv. Funct. Mater. 28 (8) (2018) 1705838.
[20] Q. Pang, X. Liang, A. Shyamsunder, L.F. Nazar, An in vivo formed solid electrolyte surface layer enables stable plating of Li metal, Joule 1 (4) (2017) 871–886.
[21] X. Ma, Z. Liu, H. Chen, Facile and scalable electrodeposition of copper current collectors for high-performance Li-metal batteries, Nano Energy 59 (2019) 500–507.
[22] L. Yu, N.L. Canfield, S. Chen, H. Lee, X. Ren, M.H. Engelhard, Q. Li, J. Liu, W. Xu, J.-G. Zhang, Enhanced stability of lithium metal anode by using a 3d porous nickel substrate, ChemElectroChem 5 (5) (2018) 761–769.
[23] L. Li, S. Basu, Y. Wang, Z. Chen, P. Hundekar, B. Wang, J. Shi, Y. Shi, S. Narayanan, N. Koratkar, Self-heating–induced healing of lithium dendrites, Science 359 (6383) (2018) 1513–1516.
[24] D.H.S. Tan, A. Banerjee, Z. Chen, Y.S. Meng, From nanoscale Interface characterization to sustainable energy storage using all-solid-state batteries, Nat. Nanotechnol. 15 (3) (2020) 170–180.
[25] P. Tan, Z. Wei, W. Shyy, T. Zhao, Prediction of the theoretical capacity of non-aqueous lithium-air batteries, Appl. Energy 109 (2013) 275–282.
[26] M. Wakihara, Recent developments in lithium ion batteries, Mater. Sci. Eng. R: Rep. 33 (4) (2001) 109–134.
[27] Z. Zhao, J. Huang, Z. Peng, Achilles' heel of lithium–air batteries: lithium carbonate, Angew. Chem. Int. Ed. 57 (15) (2018) 3874–3886.
[28] G. Girishkumar, B. McCloskey, A.C. Luntz, S. Swanson, W. Wilcke, Lithium—air battery: promise and challenges, J. Phys. Chem. Lett. 1 (14) (2010) 2193–2203.
[29] K.G. Gallagher, S. Goebel, T. Greszler, M. Mathias, W. Oelerich, D. Eroglu, V. Srinivasan, Quantifying the promise of lithium–air batteries for electric vehicles, Energy Environ. Sci. 7 (5) (2014) 1555–1563.
[30] N. Imanishi, O. Yamamoto, Perspectives and challenges of rechargeable lithium–air batteries, Mater. Today Adv. 4 (2019), 100031.
[31] L. Johnson, C. Li, Z. Liu, Y. Chen, S.A. Freunberger, P.C. Ashok, B.B. Praveen, K. Dholakia, J.-M. Tarascon, P.G. Bruce, The role of Lio2 solubility in O2 reduction in aprotic solvents and its consequences for Li–O2 batteries, Nat. Chem. 6 (12) (2014) 1091.
[32] D. Geng, N. Ding, T.S.A. Hor, S.W. Chien, Z. Liu, D. Wuu, X. Sun, Y. Zong, From lithium-oxygen to lithium-air batteries: challenges and opportunities, Adv. Energy Mater. 6 (9) (2016), 1502164.
[33] D.-J. Lee, D.-M. Im, Y.-g. Ryu, V. Roev, S.-b. Ma, M.-S. Park, Lithium Air Battery Module, Google Patents, 2016.
[34] A. Burke, M. Miller, Zero-Emission Medium-and Heavy-Duty Truck Technology, Markets, and Policy assessments for California, The University of California Institute of Transportation Studies, 2020.
[35] NHTSA, Automated Vehicles for Safety. Benefits of Automation, 2019, Available from: https://www.nhtsa.gov/technology-innovation/automated-vehicles. (Accessed 12 September 2020).
[36] SAE, Taxonomy and Definitions for Terms Related to Driving Automation Systems for on-Road Motor Vehicles - J3016_201806, On-Road Automated Driving (ORAD) Committee, SAE International, 2018, p. 35.

[37] G. Leich, J. Bischoff, Should autonomous shared taxis replace buses? A simulation study, Transp. Res. Proc. 41 (2019) 450–460.
[38] R. Abe, Introducing autonomous buses and taxis: quantifying the potential benefits in Japanese transportation systems, Transp. Res. A Policy Pract. 126 (2019) 94–113.
[39] S. Shaheen, A. Cohen, Is it time for a public transit renaissance?: navigating travel behavior, technology, and business model shifts in a brave new world, J. Public Transp. 21 (1) (2018) 8.
[40] J. Hirsch, Analysts: Trucking Ripe with Opportunities for Autonomous Technology, Business-Technology, 2020. Available from: https://www.ttnews.com/articles/analysts-autonomous-technology-make-headway-trucking-first. (Accessed 1 September 2020).
[41] G.H. Aisha Chottani, J. Murnane, F. Neuhaus, Distraction or Disruption? Autonomous Trucks Gain Ground in Us Logistics, 2018, Available from: https://www.mckinsey.com/industries/travel-logistics-and-transport-infrastructure/our-insights/distraction-or-disruption-autonomous-trucks-gain-ground-in-us-logistics. (Accessed 1 September 2020).
[42] McKinsey&Company, Autonomous Driving - Projected Impact, 2020, Available from: https://www.mckinsey.com/features/mckinsey-center-for-future-mobility/overview/autonomous-driving. (Accessed 19 September 2020).
[43] J. Hirsch, Waymo Begins Testing Autonomous Heavy-Duty Trucks in Texas. Trucking Technology - Autonomous Vehicles, 2020, Available from: https://www.trucks.com/2020/08/25/waymo-autonomous-truck-tests/. (Accessed 18 September 2020).

Index

Note: Page numbers followed by *f* indicate figures and *t* indicate tables.

C

D

R

S

T

U

V

W

FSC
www.fsc.org
MIX
Papier | Fördert
gute Waldnutzung
FSC® C083411

Kolibri 360 GmbH
Ferdinand-Jühlke-Straße 7
99095 Erfurt, Deutschland
produktsicherheit@kolibri360.de